Answers to End-of-Chapter
Study Questions for

Oceanography
An Invitation to Marine Science

Fifth Edition

Tom Garrison
Orange Coast College

THOMSON

BROOKS/COLE

Australia • Canada • Mexico • Singapore • Spain • United Kingdom • United States

Printed in the United States of America
1 2 3 4 5 6 7 08 07 06 05 04

Printer: Thomson/West

ISBN: 0-534-40891-5

For more information about our products, contact us at:
Thomson Learning Academic Resource Center
1-800-423-0563

For permission to use material from this text or product, submit a request online at
http://www.thomsonrights.com.
Any additional questions about permissions can be submitted by email to
thomsonrights@thomson.com.

Thomson Brooks/Cole
10 Davis Drive
Belmont, CA 94002-3098
USA

Asia
Thomson Learning
5 Shenton Way #01-01
UIC Building
Singapore 068808

Australia/New Zealand
Thomson Learning
102 Dodds Street
Southbank, Victoria 3006
Australia

Canada
Nelson
1120 Birchmount Road
Toronto, Ontario M1K 5G4
Canada

Europe/Middle East/
South Africa
Thomson Learning
High Holborn House
50/51 Bedford Row
London WC1R 4LR
United Kingdom

Latin America
Thomson Learning
Seneca, 53
Colonia Polanco
11560 Mexico D.F.
Mexico

Spain/Portugal
Paraninfo
Calle/Magallanes, 25
28015 Madrid, Spain

TABLE OF CONTENTS

CHAPTER 1

KNOWING THE OCEAN WORLD

Reviewing What You've Learned

1. Which hemisphere contains the greatest percentage of ocean? Is most of Earth's water in the ocean?

The Northern Hemisphere is 60.7% sea and 39.3% land; the Southern Hemisphere 80.9% sea and 19.1% land. Figure 1.2 shows the distribution of ocean and land in both hemispheres.

The ocean accounts for only slightly more than 0.02% of Earth's mass, or 0.13% of its volume. Indeed, most of Earth's water still rests deep in the rocks of the crust and mantle. At the Earth's surface, however, the ocean contains the bulk of the water -- fresh water lakes and rivers and polar icecaps account for only about 3% of the total. Although most of the ocean was in place about 4 billion years ago, ocean formation continues very slowly even today. About 0.1 cubic kilometer (0.025 cubic mile) of new water is added to the ocean each year, mostly as steam flowing from volcanic vents.

2. Which is greater: the average depth of the ocean or the average height of the continents above sea level?

The average land elevation is 840 meters (2,772 feet), but the average ocean depth is 4½ times greater at about 3,796 meters (12,451 feet).

3. Can the scientific method be applied to speculations about the natural world that are not subject to test or observation?

Science is a systematic process of asking questions about the observable world, and testing the answers to those questions. Science interprets raw information by constructing a general explanation with which the information is compatible. Scientific thought progresses as a continuing chain of questioning, testing, and matching theories to observations. A theory is strengthened if new facts support it. If not, the theory is modified or a new explanation is sought. The power of science lies in the ability of the process to operate *in reverse*; that is, in the use of a theory or law to make predictions and anticipate new facts to be observed.

The scientific method is the orderly process by which theories are verified or rejected. It is based on the assumption that nature "plays fair" -- that the answers to our questions about nature are ultimately knowable as our powers of questioning and observing improve. Theories that are not subject to questioning and observing, to test and observation, are thus not subject to the scientific method.

4. Why do we say there's one world ocean? What about the Pacific and Atlantic Oceans, the Baltic and Mediterranean Seas?

There are few dependable natural divisions, and water mixes freely within basins. There is only one great mass of water. The Pacific and Atlantic Oceans, the Mediterranean and Baltic Seas, so named for our convenience, are in reality only temporary features of a single world ocean. As we will see in a later chapter, the ocean floor moves, and continents shift positions, continuously altering the shape and position of the ocean's basins.

5. What were the stimuli to Polynesian colonization? How were the long voyages accomplished? How were Polynesian voyages different from (and similar to) those of the Vikings?

The ancestors of the Polynesians spread eastward from Southeast Asia or Indonesia in the distant past. Although experts vary in their estimates, there is some consensus that by 30,000 years ago New Guinea was populated by these wanderers and by 20,000 years ago the Philippines were occupied. By around 500 B.C. the so-called cradle of Polynesia -- Tonga, Samoa, the Marquesas and the Society islands -- was settled and the Polynesian cultures formed.

For a long and evidently prosperous period the Polynesians spread from island to island until the easily accessible islands had been colonized. Eventually, however, overpopulation and depletion of resources became a problem. Politics, intertribal tensions, and religious strife shook their society. When tensions reached the breaking point, groups of people scattered in all directions from the Marquesas and Society Islands during a period of explosive dispersion. Between 300 and 600 A.D. Polynesians successfully colonized nearly every inhabitable island within the vast triangular area shown in Figure 1.9. Easter Island was found against prevailing winds and currents, and the remote islands of Hawaii were discovered and occupied. These were among the last places on Earth to be populated.

Large dual-hulled sailing ships, some capable of transporting up to 100 people, were designed and built for the voyages. New navigation techniques were perfected that depended on the positions of stars barely visible to the north. New ways of storing food, water, and seeds were devised. In that anxious time the Polynesians honed and perfected their seafaring knowledge. To a skilled navigator a change in the rhythmic set of waves against the hull could indicate an island out of sight over the horizon. The flight tracks of birds at dusk could suggest the direction of land. The positions of the stars told stories, as did the distant clouds over an unseen island. The smell of the water, or its temperature, or salinity, or color, conveyed information, as did the direction of the wind relative to the sun, and the type of marine life clustering near the boat. The sunrise colors, sunset colors, the hue of the moon -- every nuance had meaning, every detail had been passed in ritual from father to son. The greatest Polynesian minds were navigators, and reaching Hawaii was their greatest achievement.

The Vikings, on the other hand, began their travels as raiders. Their initial voyages were motivated by the twin desires of acquiring valuable objects and avoiding the rigors of Scandinavian winters. Leadership within groups was by clan relation or success in battle, and voyages were planned by anticipated results rather than for religious, sociological, or exploratory reasons. Rapid advances in shipbuilding and navigation made raids almost routine (much to the disgust of the residents of coastal Europe). When better defenses made raiding less attractive, the Vikings looked westward. The discovery of Vinland (North America) came at a propitious time -- being a Viking was not a profession with a bright future, but a career in real estate advertising and sales definitely had promise.

Both groups knew enough about the ocean to travel dependably over vast distances. Of course, they never met. Their motivations were different, but their techniques were roughly similar.

6. What were the main stimuli to European voyages of exploration during the Age of Discovery? Why did it end?

There were two main stimuli: (1) encouragement of trade, and (2) military one-upsmanship.

Trade between east and west had long been dependent on arduous and insecure desert caravan routes through the central Asian and Arabian deserts. This commerce was cut off in 1453 when the Turks captured Constantinople, and an alternate ocean route was needed. As we have seen, Prince Henry the Navigator (of Portugal) thought ocean exploration held the key to great wealth and successful trade. Henry's explorers pushed south into the unknown and opened the West coast of Africa to commerce. He sent out small, maneuverable ships designed for voyages of discovery and manned by well-trained crews. Henry's students knew the Earth was round but, thanks to the errors of Claudius Ptolemy, were wrong in their understanding of its true size.

Christopher Columbus was familiar with Prince Henry's work, and "discovered" the New World quite by accident while on a mission to encourage trade. His intention was to pioneer a sea route to the rich and fabled lands of the east made famous more than 200 years earlier in the overland travels of Marco Polo. As "Admiral of the Ocean Sea," Columbus was to have a financial interest in the trade routes he blazed. As we saw, Columbus never appreciated the fact that he had found a new continent. He went to his grave confident that he had found islands just off the coast of Asia.

Charts that included the properly-identified New World inspired Ferdinand Magellan, a Portuguese navigator in the service of Spain, to believe that he could open a westerly trade route to the Orient. In the Philippines, Magellan was killed and his crew decided to continue sailing west around the world. Only 18 of the original 250 men survived, returning to Spain three years after they set out. But they had proved it was possible to circumnavigate the globe.

The seeds of colonial expansion had been planted. Later, the empires of Spain, Holland, Britain, and France pushed into the distant oceanic reaches in search of lands to claim. Military strength might depend on good charts, knowledge of safe harbors in which to take on provisions, and friendly relations with the locals. Exploration was undertaken to insure these things.

But that gets ahead of the story. The Magellan expedition's return to Spain in 1522 -- the end of the first circumnavigation -- technically marks the end of the first age of European discovery.

7. What were the oceanographic contributions of Prince Henry the Navigator, Benjamin Franklin, Matthew Maury, and Charles Darwin?

Prince Henry the Navigator: Third son of the royal family of Portugal, Prince Henry established a center at Sagres for the study of marine science and navigation ". . . through all the watery roads." Although he personally was not well traveled (he went to sea only twice in his life), captains under his patronage explored from 1451 to 1470, compiling detailed charts wherever they went. Henry's explorers pushed south into the unknown and opened the west coast of Africa to commerce. He sent out small, maneuverable ships designed for voyages of discovery and manned by well-trained crews. Henry's students knew the Earth was round (but because of the errors publicized by Claudius Ptolemy they were wrong in their estimation of its size).

Benjamin Franklin: While postmaster of the colonies, Franklin had noticed the peculiar fact that the fastest ships were not always the fastest ships; that is, hull speed did not always correlate with out-and-return time on the European run. Franklin's cousin, a Nantucket merchant named Tim Folger, noted Franklin's puzzlement and provided him with a rough chart of the "Gulph Stream" that he (Folger) had worked out. By staying within the stream on the outbound leg and adding its speed to their own, and by avoiding it on their return, captains could traverse the Atlantic much more quickly. It was Franklin who published, in 1769, the first chart of any current.

Matthew Maury: Maury who was the first person to sense the worldwide pattern of surface winds and currents. Based on his analysis, he produced a set of directions for sailing great distances more efficiently. Maury's sailing directions quickly attracted worldwide notice: He had shortened the passage for vessels traveling from the American East Coast to Rio de Janeiro by 10 days, and to Australia by 20. His work became famous in 1849 during the California gold rush—his directions made it possible to save 30 days on the voyage around Cape Horn to California. His crowning achievement, *The Physical Geography of the Seas*, a book explaining his discoveries, was published in 1855. Maury, considered by many to be the father of physical oceanography, was perhaps the first man to undertake the systematic study of the ocean as a full-time occupation.

Charles Darwin: Charles Darwin's first contributions to science were in the field of marine biology. In his first major work, *Structure and Distribution of Coral Reefs* (1842), Darwin correctly argued that atolls and reefs such as those he observed could only have resulted from subsidence of the seafloor, with corals growing upward as their bases dropped. His monographs on barnacle biology, volcanic islands, and fossils are overshadowed by his remarkable work *On the Origin of Species*, published in 1859.

8. How can you find your approximate latitude and longitude at sea?

Latitude is relatively simple to determine. In the northern hemisphere, take a simple protractor and measure the angle between the horizon, your eye, and the north polar star. The protractor reads approximately in degrees of latitude. See the appendix on time and navigation at the back of your book for more information.

To find longitude, you'll need an accurate clock. First, determine local noon by observing the path of a shadow of a vertical shaft -- it is shortest at noon -- and set your clock accordingly. After traveling some distance to the west, you will notice that noon according to your *clock* no longer marks the time when the shadow of the *shaft* is shortest at your new location. If "clock" noon occurs three hours before "shaft" noon, you can do some simple math to see how far west of your starting point you have come. The Earth turns toward the east, making one rotation of 360 degrees in 24 hours, so its rotation rate is 15° per hour (360°/ 24 hours = 15°/hour). The three hours' difference between "clock" noon and "shaft" noon puts you 45° west of your point of origin (3 x 15°). The more accurate the clock (and the measurement of the shaft's shadow), the more accurate your estimate of westward position.

In 1998 the cost of small, hand-held GPS (global positioning system) devices capable of reading one's position from an orbiting constellation of satellites fell below $100. Weekend sailors can find their way into harbor entrances in thick fog, and hikers can return from the forest to their sport utility vehicles, with an ease and accuracy that would have sent Cook, Eratosthenes, or Magellan into paroxysms of disbelief.

9. What were the goals and results of The United States Exploring Expedition? What U.S. institution greatly benefited from its efforts?

The United States Exploring Expedition, launched in 1838, was primarily a naval expedition. Its goals included showing the flag, whale scouting, mineral gathering, charting, observing, and pure exploration. The scientific team explored and charted a large sector of the east Antarctic coast, and made observations that confirmed the landmass as a continent. A map of the Oregon territory produced in 1841, one of 241 maps and charts drawn by members of the expedition, proved especially valuable when connected to the map of the Rocky Mountains prepared the following year by Captain John C. Fremont. Hawaii was thoroughly explored, and Wilkes led an ascent of Mauna Loa, one of the two highest peaks of Hawaii's largest island. One unusual goal was to disprove a peculiar theory that the Earth was hollow and could be entered through huge holes at either pole. The work of the scientists aboard the flagship *Vincennes* and the expedition's five other vessels helped to establish the natural sciences as reputable professions in America. (No evidence of polar holes was found!)

The expedition returned with many scientific specimens and artifacts, which formed the nucleus of the collection of the newly established Smithsonian Institution in Washington, D.C.

10. What was the first purely scientific oceanographic expedition, and what were some of its accomplishments? What contributions did the earlier, hybrid expeditions make?

The first sailing expedition devoted completely to marine science was the HMS *Challenger* expedition, conceived by a professor of natural history at Scotland's University of Edinburgh, Charles Wyville Thomson, and his Canadian-born student of natural history, John Murray. Stimulated by their own curiosity and by the inspiration of Charles Darwin's 1831-1836 voyage in HMS *Beagle*, they convinced the Royal Society and the British Government to provide a Royal Navy ship and trained crew for a "prolonged and arduous voyage of exploration across the oceans of the world." As you may remember reading, Thomson and Murray even coined a word for their enterprise: Oceanography. The government and the Royal Society agreed to the endeavor provided a proportion of any financial gain from discoveries was handed over to the Crown. This arranged, the scientists made their plans.

HMS *Challenger*, a 2,306 ton steam corvette, set sail on 7 December 1872 on a four-year voyage that took them around the world and covered 127,600 kilometers (79,300 nautical miles). Although the Captain was a Royal Naval officer, the six-man scientific staff directed the course of the voyage. The crew made an extraordinary 492 deep soundings with mechanical grabs and nets at 362 stations (including 133 dredgings). With each hoist animals new to science were strewn on the deck; in all, staff biologists discovered 4,717 new species! The scientists also took salinity, temperature, and water density measurements during these soundings. Each reading contributed to a growing picture of the physical structure of the deep ocean. They completed at least 151 open water trawls, and stored 77 samples of seawater for detailed analysis ashore. The expedition collected new information on ocean currents, meteorology, and the distribution of sediments; the locations and profiles of coral reefs were charted. Thousands of pounds of specimens were brought to British museums for study. Manganese nodules, brown lumps of mineral-rich sediments, were discovered on the seabed, sparking interest in deep sea mining.

This first pure oceanographic investigation was an unqualified success. The discovery of life in the depths of the oceans stimulated the new science of marine biology. The scope, accuracy, thoroughness, and attractive presentation of the researchers' written reports made this expedition a high point in scientific publication. The *Challenger Report*, the record of the expedition, was published between 1880 and 1895 by Sir John Murray in a well-written and magnificently illustrated 50-volume set; it is still used today. Indeed, it was the 50 volume *Report*, rather than the cruise, that provided the foundation for the new science of oceanography. The *Challenger* expedition remains history's longest continuous scientific oceanographic expedition.

Scientific oceanography, however, began earlier. Perhaps the first scientific voyage, a hybrid scientific/military expedition, was the 1768- 1771 voyage of James Cook of the British Royal Navy. An intelligent and patient leader, Cook was also a skillful navigator, cartographer, writer, artist, diplomat, sailor, scientist, and dietician. The primary reason for the voyage was to assert the British presence in the South Seas, but the expedition had numerous scientific goals as well. First Cook conveyed several members of The Royal Society (a scientific research group) to Tahiti to observe the transit of Venus across the disc of the sun. These measurements verified calculations of planetary orbits made earlier by Edmund Halley (later of comet fame) and others. Then Cook turned south into unknown territory to search for a hypothetical southern continent which philosophers believed must exist to balance the landmass of the northern hemisphere. Cook and his men found and charted New Zealand, mapped Australia's Great Barrier Reef, marked the positions of tens of small islands, made notes on the natural history and human inhabitation of these distant places, and initiated friendly relations with many chiefs. Cook survived an epidemic of dysentery contracted by ship's company while ashore in Batavia (Djakarta), and sailed home to England around the world in 1771. Because of his insistence on cleanliness and ventilation, and because his provisions included cress, sauerkraut, and citrus extracts his sailors avoided scurvy, a Vitamin-C deficiency disease which for centuries had decimated crews on long voyages. Cook made two other important voyages of discovery and observation.

Cook deserves to be considered a scientist as well as an explorer because of his accuracy, thoroughness, and the completeness in his descriptions. He and the scientists aboard took samples of marine life, land plants and animals, the ocean floor, and geological formations, and reported their characteristics in their logbooks and journals. His navigation was outstanding, and his charts of the Pacific were accurate enough to be used by the allied powers in Second World War invasions of the Pacific islands. He drew accurate conclusions, did not exaggerate his findings in his reports, and opened friendly diplomatic relations with many native populations. Cook recorded and successfully interpreted events in natural history, anthropology, and oceanography. Unlike most captains of his day he cared for his men. He was a thoughtful and clear writer. This first marine scientist peacefully changed the map of the world more than any explorer or scientist in history.

Thinking Critically

1. How could you convince a 10-year old that the Earth is round? What evidence would the child offer that it's flat? How can you counter those objections?

You might point out the roundness of the sun and the moon, show him or her the spherical shape of a few selected planets when seen through a telescope, and then suggest that, like a library globe, the Earth is of the same type. Point to the shadow of the Earth on the moon during a lunar eclipse. Watch ships go hull down, then mast down, then flag down as they leave a shore.

The child would explain the obvious foolishness of such a view. If the Earth were round, people on the bottom would fall off.

You would then pull out your copy of Newton's *Principia Mathematica* and turn to the diagrams showing the centripetal nature of gravitational force.

The child would be impressed, but disbelieving. Try again when he or she is 18.

2. How did the Library at Alexandria contribute to the development of marine science? What happened to most of the information accumulated there? Why do you suppose the residents of Alexandria became hostile to the librarians and the many achievements of the Library?

The great Library at Alexandria constituted history's greatest accumulation of ancient writings. As you read, the characteristics of nations, trade, natural wonders, artistic achievements, tourist sights, investment opportunities, and other items of interest to seafarers were catalogued and filed in its stacks. Manuscripts describing the Mediterranean coast were of great interest.

Traders quickly realized the competitive benefit of this information. Knowledge of where a cargo of olive oil could be sold at the greatest profit, or where the market for finished cloth was most lucrative, or where raw materials for metalworking could be obtained at low cost, was of enormous competitive value. Here perhaps was the first instance of cooperation between a university and the commercial community, a partnership that has paid dividends for science and business ever since.

After their market research was completed, it is not difficult to imagine seafarers lingering at the Library to satisfy their curiosity about non-commercial topics. And there would have been much to learn! In addition to Eratosthenes' discovery of the size of the Earth, Euclid systematized geometry; the astronomer Aristarchus of Samos argued that the Earth is one of the planets and that all planets orbit the sun; Dionysius of Thrace defined and codified the parts of speech (noun, verb, etc.) common to all languages; Herophilus, a physiologist, established the brain was the seat of intelligence; Heron built the first steam engines and gear trains; Archimedes discovered (among many other things) the principles of buoyancy on which successful shipbuilding is based.

The last Librarian was Hypatia, the first notable woman mathematician, philosopher, and scientist. In Alexandria she was a symbol of science and knowledge, concepts the early Christians identified with pagan practices. After years of rising tensions, in 415 A.D. a mob brutally murdered her and burned the Library with all its contents. Most of the community of scholars dispersed and Alexandria ceased to be a center of learning in the ancient world.

The academic loss was incalculable, and trade suffered because ship owners no longer had a clearing house for updating the nautical charts and information upon which they had come to depend. All that remains of the

Library today is a remnant of an underground storage room. We shall never know the true extent and influence of its collection of over 700,000 irreplaceable scrolls.

Historians are divided on the reasons for the fall of the Library. But we know there is no record that any of the Library's scientists ever challenged the political, economic, religious, or social assumptions of their society. Researchers did not attempt to explain or popularize the results of their research, so residents of the city had no understanding of the momentous discoveries being made at the Library at the top of the hill. With very few exceptions, the scientists did not apply their discoveries to the benefit of mankind, and many of the intellectual discoveries had little practical application. The citizens saw no practical value to such an expensive enterprise. Religious strife added elements of hostility and instability. As Carl Sagan points out, "When, at long last, the mob came to burn the Library down, there was nobody to stop them."[1]

3. How did Eratosthenes calculate the approximate size of the Earth? Which of his assumptions was the "shakiest"?

Eratosthenes and his educated contemporaries were nearly certain the Earth was a sphere. To the evidence listed in Question #1 above, he added the observation that in Alexandria, a vertical pole cast a slight shadow at noon on the longest day of the year (when the sun is as high in the sky as possible). Eratosthenes had heard from travelers returning from Syene (now Aswan, site of the great Nile dam) that at noon on the longest day of the year the sun shone *directly* down onto the waters of a deep vertical well. He measured the shadow angle at Alexandria and found it to be a bit more than 7°, about one fiftieth of a circle. If the sun were directly overhead at Syene, but not directly overhead at Alexandria, then the surface of the Earth must be curved.

But what was the *circumference* of the Earth? By studying the reports of camel caravan traders, he estimated the distance from Alexandria to Syene at about 785 kilometers (491 miles). Eratosthenes now had the two pieces of information needed to derive the circumference of the Earth by geometry. Figure 2.2 shows his solution.

The "shakiest" assumption was that the sun was infinitely far away. Only in that position would descending rays approaching both Alexandria and Syene be truly parallel. We know the sun is not infinitely far away, of course, but the sun is sufficiently distant to produce rays parallel enough for Eratosthenes to generate a surprisingly accurate circumference.

---

[1] Sagan, C. 1980. Cosmos. New York: Random House.

4. If Columbus did not discover North America, then who did?

Columbus's landfalls were all south of North America. North America was itself discovered by the first humans to follow migratory game from west to east across the Bering Strait ice bridge (west of Alaska) during the last ice age. Estimates vary widely, but the date of the overland arrival of the first North Americans was probably 20,000 to 35,000 years ago. So, the "locals" discovered North America.

5. Sketch briefly the major developments in marine science since 1900. Do individuals, separate voyages, or institutions figure most prominently in this history?

A look at the part of Table 1.1 on page 33 will help. The undeniable success story of late twentieth century oceanography is the successful rise of the great research institutions with broad state and national funding. Without the cooperation of research universities and the federal government (through agencies like the National Science Foundation, the National Oceanic and Atmospheric Administration, and others), the great strides that were made in the fields of plate tectonics, atmosphere-ocean interaction, biological productivity, and ecological awareness would have been much slower in coming. Along with the Sea Grant Universities (and their equivalents in other countries), establishments like the Scripps Institution of Oceanography, the Lamont-Doherty Earth Observatory, and the Woods Hole Oceanographic Institution, with their powerful array of researchers and research tools, will define the future of oceanography.

Now oceanography from space is gaining ascendance. A brief look at the spectacular photographs and data renderings in your textbook will demonstrate the growing power of satellite oceanography.

Thinking Analytically

1. Imagine you set your watch at local noon in Kansas City on Monday, then fly to the Coast on Tuesday. You stick a pole into the ground on a sunny day at the beach, wait until its shadow is shortest, and look at your watch. The watch says 10 a.m. Are you on the East Coast or the West Coast? What is the difference in longitude? (Hint: 360° divided by 24 hours is 15°. The sun moves through the sky at a rate of 15° per hour).

Re-read the answer to Question #8 in the Review Questions above. Remember, the sun moves across the sky from east to west. If pole noon comes earlier than clock noon, you're east of clock noon, on the east coast. A two hour difference in time is equivalent to a 30° difference in longitude.

2. Look at Box 1.1, figure b. Provide a rough estimate of the latitude and longitude of your home.

This one is up to you. Ask your instructor or teaching assistant if you're in the ball park.

3. Magellan's crew kept very careful records of their circumnavigation, yet when they returned home, they were one day "off." Why? Had they gained a day, or lost a day?

Magellan's crew went around the world in a westerly direction. Earth rotates to the east. A crewman wrote (in Tim Joyner's translation): "...and I was more surprised than the others, since having always been in good health, I had every day, without interruption, written down the day that was current." Later, in Spain, he was advised that "...there was an error on our part, since as we had always sailed toward the west, folling the course of the suh, and had returned to the same place, we must have gained twenty-four hours, aa it is clear to anyone who reflects on it."

4. Replicate Eratosthenes' measurement of the diameter of Earth. Try this technique: Contact a friend who lives about 800 kilometers (500 miles) north or south of you (a distance comparable to the distance between Alexandria and Syene.) Drive a tall pole into the ground at each location. Make sure the poles are vertical (using a weight on a string). Watch around noon, and when the pole casts the shortest shadow, measure the sun's angle of inclination from the shadow cast. Can you take it from there?

Give it a try. If your friend is far enough north or south of you, and you both measure carefully, and you can get a working solution.

CHAPTER 2

ORIGINS

Reviewing What You've Learned

1. How old is Earth? On what are these estimates based?

Many lines of evidence interlock to suggest Earth is about 4.6 billion (4,600 million) years old.

Radiometric dating (described in Box 3.2 on page 66) is a powerful technique based on the discovery that unstable, naturally radioactive elements lose particles from their nuclei and ultimately change into new stable elements. The radioactive decay occurs at a predictable rate, and measuring the ratio of radioactive to stable atoms in a sample provides its age. Using radiometric dating, researchers have identified small zircon grains from western Australian sandstone that are 4.2 billion years old. The zircons were probably eroded from nearby continental rocks and deposited by rivers. (Older crust is now unidentifiable, having been altered and converted into other rocks by geological processes.)

Observing the rates of mountain-building and erosion also provides clues. If we assume the processes we observe occur now at rates similar to rates in the past, we can extrapolate suggestions of age. Even the rate at which heat leaks from within Earth can provide data.

One of the most interesting methods of age-dating searches for cosmic-ray traces in metallic meteorites that have fallen from space. It is likely these objects are remnants of the cores of failed planets, or from material ejected from Earth that ended up in our moon. The density of the traces suggests how long it has been since the objects formed.

The moon itself has provided clues. Its angular momentum and orbital shape can tell of its early formation and subsequent movements, and samples brought back by Apollo astronauts have confirmed an age similar to (but slightly younger than) Earth.

2. What is the relationship between galaxies and stars? What things make up galaxies?

A galaxy is a huge rotating aggregation of stars, dust, gas, and other debris held together by gravity. Our galaxy is named the Milky Way.

The stars that make up a galaxy are massive spheres of incandescent gases. They are usually intermingled with diffuse clouds of gas and debris laced with heavy elements. In spiral galaxies like the Milky Way, the stars are arrayed in curved arms radiating from the galactic center. Smaller galaxies may be loosely associated with a large galaxy (ours has at least two hangers-on), and clusters of ancient stars may surround the galactic nucleus like a huge halo. Our part of the Milky Way is populated with many stars, but distances within a galaxy are so huge that the star nearest the sun is about 42 trillion kilometers (26 trillion miles) away. Astronomers tell us there are perhaps 100 billion galaxies in the universe and 100 billion stars in each galaxy. Imagine more stars in the Milky Way than grains of sand on a beach!

3. What element makes up most of the detectable mass in the universe.

Hydrogen. This answer is interesting – more than 95% of all the mass in the universe appears to be hydrogen. But Earth is not made mostly of hydrogen. Do you remember how the heavy elements were made? (If not, see #6 below.)

4. Briefly trace the life of a typical star.

Operating under the influence of gravity, a diffuse mass of hydrogen in the spiral arm of a galaxy can shrink, the knot of gases becoming a protostar. The original diameter of the protostar may be many times the diameter of our solar system, but gravitational energy causes it to contract. As the protostar shrinks, the compression raises its internal temperature. When the protostar reaches a temperature of about 10 million degrees Celsius (18 million degrees Fahrenheit) nuclear fusion begins; that is, hydrogen atoms begin to fuse together to form helium, a process that liberates even more energy. This rapid release of energy, which marks the transition from protostar to star, stops the young star's shrinkage. Nuclear fusion of small atoms, not combustion, is what causes a star to shine. Light and heat, byproducts of the new star, make the galaxy a little brighter.

After fusion reactions begin, the star becomes stable, neither shrinking nor expanding and burning its hydrogen fuel at a steady rate. This stable phase does not last forever, however. After a long and productive life, the star has converted a large percentage of its hydrogen to atoms as heavy as carbon or oxygen. Very heavy elements are formed in the brief moments of a massive star's death. The explosive release of energy in a supernova is so sudden that the star is blown to bits, and its shattered mass is accelerated outward at nearly the speed of light. The explosion lasts only about 30 seconds, but in that short time the nuclear forces holding apart individual atomic nuclei are overcome and atoms heavier than iron are formed. The gold of your rings, the mercury in a thermometer, and the uranium in nuclear power plants are all created during that brief and stupendous flash. The atoms produced by the star through millions of years of orderly fusion, *and* the heavy atoms generated in a few moments of unimaginable chaos, are sprayed into space. Every chemical element heavier than hydrogen -- most of the atoms that makes up the planets, the oceans, and living creatures -- was manufactured by the stars.

5. How are planets and stars related?

Both stars and planets condense from clouds of galactic dust and gas. Planets typically shine by reflected light, while stars generate their own light. Stars tend to be more massive than planets, and form the center of mass of rotating star-planet systems. Planets containing heavy elements could not exist until at least one generation of stars had constructed those elements.

6. How are light elements converted into heavy ones?

Stars are responsible for this bit of alchemy. During long and productive lives, stars convert a large percentage of their hydrogen to atoms as heavy as carbon or oxygen. These are driven into space during periodic expansions of the star's outer envelope. Very heavy elements are formed in the brief moments of a massive star's death. The explosive release of energy in a supernova is so sudden that the star is blown to bits, and its shattered mass is accelerated outward at nearly the speed of light. The explosion lasts only about 30 seconds, but in that short time the nuclear forces holding apart individual atomic nuclei are overcome and atoms heavier than iron are formed. The gold of your rings, the mercury in a thermometer, and the uranium in nuclear power plants are all created during that brief and stupendous flash.

7. Will all stars end their lives as supernovas? What happens to the heavy elements made by small stars?

The life history of a star described in Question #4 depends on a high initial mass. When a medium-mass star begins to consume atoms heavier than hydrogen and helium -- atoms as heavy as carbon or oxygen -- its energy output slowly rises and its body swells to a stage aptly named "red giant" by astronomers. The dying giant slowly pulsates, throwing off concentric shells of light gas enriched with these heavy elements. But there is no supernova explosion. Most of the harvest of carbon and oxygen is forever trapped in the cooling ember at the star's heart.

8. What is "density stratification?"

The formation of layers in a material, with each deeper layer being denser (weighing more per unit of volume) than the layer above. Earth itself is density stratified. The atmosphere and ocean are usually density stratified.

9. Are the ocean and present atmosphere "leftovers" from the original atmosphere of Earth?

Yes and no -- ancient materials have been combined in new ways as time has passed.

In a sense, there have been three atmospheres: the original atmosphere blown away by the ignition shock of the sun, the reducing atmosphere that outgassed from within the Earth, and the oxidizing atmosphere that has resulted from the work of photosynthesizing plants and plant-like organisms.

The volcanic venting of volatile substances including water vapor -- outgassing -- gave rise to the present ocean. As hot water vapor rose, it condensed into clouds in the cool upper atmosphere. Recent research suggests that millions of tiny icy comets colliding with the Earth may also have contributed to the accumulating mass of water vapor, this ocean-to-be.

10. What is biosynthesis? Where do researchers think it might have occurred on our planet? Could it happen again today?

Biosynthesis is the term given to the early evolution of living organisms from the simple organic building blocks present on and in the early Earth.

Two potential sites for biosynthesis have been suggested. In the 1950s, biologists suggested that life may have originated in shallow tidal pools at the ocean's edge. Evaporation of water from these pools would have concentrated the amino acid and nucleotide building blocks into a rich organic "soup." Grains of sand or tiny bubbles would have provided handy surfaces on which larger chemical combinations could be assembled. Sunlight would have supplied the energy for these reactions. Accumulating in protected pools, reacting aggregates could have become progressively more complex, eventually evolving into biochemical systems capable of reproducing. Unfortunately for this hypothesis, the sunlight needed to energize the reactions might not have been present; indeed, strong sunlight has harmful effects on unprotected large molecules. Recently, planetary scientists have suggested that the sun was faint in its youth. It put out so little heat that the ocean may have been frozen to a depth of around 300 meters (1,000 feet). The ice would have formed a blanket that kept most of the ocean fluid and relatively warm. Periodic fiery impacts by asteroids could have thawed the ice, but between batterings it would have reformed. In 1994, chemists Jeffrey Bada and Stanley Miller suggested that organic material may have formed and then been trapped beneath the ice, protected from the atmosphere which contained chemical compounds capable of shattering the complex molecules. The first self-sustaining molecules might have arisen deep below the layers of surface ice, on clays or pyrite crystals near warm hydrothermal vents on the ocean floor. Most researchers agree, however, that whatever the details, some quiet corner of the ocean was the likely place where life began. Life probably originated only a few hundred million years after a stable ocean formed. Life and the Earth have grown old together; each has greatly influenced the other.

A similar biosynthesis could not occur today. Living things have changed the conditions in the ocean and atmosphere, and those changes are not consistent with any new origin of life. For one thing, green plants have filled the atmosphere with oxygen, a compound that can disrupt any unprotected large molecule. For another, some of this oxygen (as ozone) now blocks much of the ultraviolet radiation from reaching the surface of the ocean. And finally, the many tiny organisms present today would gladly scavenge any large organic molecules as food.

Thinking Critically

1. Marine biologists sometimes say that all life-forms on Earth, even desert lizards and alpine plants, are marine. Why?

The fact that all life, from a jellyfish to a dusty desert weed, depends on saline water within its cells to dissolve and transport chemicals is certainly significant. It strongly suggests that simple, self-replicating -- living -- molecules arose somewhere in the early ocean. By extension, one might argue that since all forms of life on this planet contain an analog of seawater, all life is (in a sense) marine.

2. Where did the Earth's surface water come from?

The Earth's surface was so hot that no water could settle there, and no sunlight could penetrate the thick clouds. After millions of years the upper clouds cooled enough for some of the outgassed water to form droplets. Hot rains fell toward the Earth only to boil back into the clouds again. As the surface became cooler, water collected in basins and dissolved minerals from the rocks. Some of the water evaporated, cooled, and fell again. The world ocean was gradually accumulating.

These heavy rains may have lasted for 10 million years. Large amounts of water vapor and other gases continued to escape through volcanic vents during that time and for millions of years thereafter. The ocean grew deeper.

3. Do you think water planets are common in the galaxy? What about planets in general?

I wouldn't expect to encounter many water planets.

For starters, let's look at stars. Most stars visible to us are members of multiple-star systems. If the Earth were in orbit around a typical multiple-star system, we would be close to at least one of the host stars at certain places in our orbit, and too far away at others. Also, not all stars -- in single or multiple systems -- are as stable and steady in energy output as our sun. If we were in orbit around a star that grew hotter and cooler at intervals, our situation would be radically different than it is at the moment.

Next, let's look at orbital characteristics. Our Earth is in a nearly circular orbit at just the right distance from the sun to allow liquid water to exist over most of the surface through most of the year.

Next, consider our planet's cargo of elements. We picked these up during the accretion phase. At our area of orbit there was an unusually large amount of water (or chemical materials that would led to the formation of water).

So, with a stable star, a pleasant circular orbit that is well placed, and suitable and abundant raw materials, we are a water planet. This marvelous combination is probably not found in many places in the galaxy.

As for planets in general, there are at least 125 known as I write this, and planets are being discovered at a rate exceeding two per month. Nearly all of these are large gas-giant (Jupiter-like) beasts, bodies large enough to disrupt the motion of their parent stars through space (see Question #3 in "Thinking Analytically," below). Very few candidates for "Earth-like" have been found.

4. How might Earth be different if an ocean had not formed on its surface?

Because the presence of the moderating influence of large quantities of liquid water on Earth's surface, atmospheric temperatures would be much higher in summer, and much lower in winter. The composition of the atmosphere would be much different, a combination of methane, ammonia, carbon dioxide, and other gases. Because no life (at least no life as we know it) would have originated on the surface, the transformation to an oxygen-rich atmosphere would not have occurred. And we certainly wouldn't be around to ask questions and study the planet.

5. How do we know what happened so long ago?

Age estimates are derived from interlocking data obtained by many researchers using different sources. One source is meteorites, chunks of rock and metal formed at about the same time as the sun and planets and out of the same cloud. Some meteorites contain gases thought to be remnants of the original solar nebula. Many have fallen to Earth in comparatively recent times. We know from signs of radiation within these objects how long it has been since they were formed. That information, combined with the rate of radioactive decay of unstable atoms in meteorites, moon rocks, and the oldest rocks on the Earth, allows astronomers to make a reasonably accurate estimate of how long ago these objects formed.

In other chapters you'll learn about other methods of determining age. A few of these are radiocarbon dating, analysis of sediment layers, and inspection of rocks and minerals in Earth's crust.

6. Considering what must happen to form them, do you think ocean worlds are relatively abundant in the galaxy? Why or why not?

This is an inadvertent repeat of Question #3, above. That answer will work here, too.

7. Could water planets form around multiple star systems? Stars with greatly variable output?

One could imagine a water planet's formation by the same processes that gave rise to Earth, but its subsequent life near a multiple-star system or greatly variable star would be much less orderly.

A planetary orbit around a multiple-star system would take one of two paths: Orbiting the stars' combined center of mass at a great distance in a nearly circular orbit, or dodging through the inner chaos of the orbiting stars themselves. In the first instance, any stable orbit would be very far away from the stars, and thus would receive very little energy from them. Any ocean would be frozen solid. In the second instance, the planet itself – let along any ocean – would have a very brief lifespan! It would either crash into one of the stars or be flung out of the system altogether. Interestingly, most of the stars in our galaxy are members of multiple-star systems.

Variable stars are also very common – our own sun is a variable star. We have survived periods of increasing or decreasing solar output over the last few billion years, but the variations are only a very few percentage points. Any star with a significant variability would alternative boil or freeze the surface of a water planet in an orbit comparable to Earth's

Thinking Analytically

1. A light-year is the distance light can travel in one year. Light travels at 300,000 kilometers (186,000 miles) per second. Commercial television broadcasting began in 1939. Television signals travel at the speed of light. How far away would a space probe have to be before it could no longer detect those signals?

A light-year is about 6 trillion miles. As I write this, it has been 65 years since commercial television began, so the sphere that includes human television broadcasting (centered, of course, on the Earth) is 780 trillion miles in diameter. A probe half that distance away (the radius, 780/2 = 290 trillion miles) would be out of the range of television signals. You might be interested to know that about 200 stars are now within that radius, and, with the right equipment, inhabitants of any planets orbiting them would be able to detect our programming.

2. Density is mass per unit volume. Granite rock weighs about 2.6 g/cm³, water weighs about 1.0 g/cm³. Knowing their sizes, how might you determine whether Europa or Ganymede is hiding a large liquid water ocean beneath an icy crust?

You would start by watching the orbit of the moons around their primary (Jupiter in this case). If you know the moons' sizes, and their speeds and diameters of their orbital motions, you can calculate their average density. Now place a spacecraft in an elliptical orbit around each moon. The craft will speed up as it approaches the moon, and decelerate as it moves away. Are the rates of acceleration and deceleration what you would expect if the moon were homogeneous? If not, what composition of the outer layers (in contrast to the inner layers) would you postulate to account for the variations observed? Does water, with its relatively low density of 1.0 g/cm³ look like a likely candidate to explain the amount of gravitational pull?

3. Can you think of any way an astronomer could detect a large planet orbiting a star without actually seeing the planet? (Hint: How would the star move as the planet orbits it?)

On a dark night, put on a lighted hat and walk on smooth ground straight away from an observer. He sees the light getting dimmer, but the light does not change its position – if you're walking north, the light always heads north.

Now return to your starting point and pick up a short rope tied to a concrete block. Swing the rope around overhead as you walk away. Now the observer will see the light wobble back and forth as you walk north – you're once-steady departure is moving side-to-side because the rotating mass you're holding is altering your path in a series of oscillations as you proceed. The light on your head moves left and right and back again.

Planets don't shine with their own light; stars do. If a planet is too dim to see, you can still deduce its presence by its effect on its star. If the planet is big enough and close enough, the star will wobble from side to side as it moves through space. This method has been used to find about two-thirds of the extrasolar planets now known.

CHAPTER 3

EARTH STRUCTURE AND PLATE TECTONICS

Reviewing What You've Learned

1. Name Earth's interior layers and their properties. Which classification method (chemical or physical) is more useful in explaining plate tectonics?

Earth's internal layers may be classified by chemical composition or by physical properties.

Chemical composition: The thin oceanic crust is primarily basalt, a heavy dark colored rock composed mostly of oxygen, silicon, magnesium, and iron. By contrast, the most common material in the thicker continental crust is granite, a familiar speckled rock composed mainly of oxygen, silicon, and aluminum. The mantle, the layer beneath the crust, is thought to consist mainly of oxygen, magnesium, and silicon. The outer and inner cores, which consist mainly of iron, lie beneath the mantle at the Earth's center.

Physical properties: Different conditions of temperature and pressure prevail at different depths, and these conditions influence the physical properties of the materials subjected to them. The behavior of a rock is determined by three factors: temperature, pressure, and the rate at which a deforming force (stress) is applied. This behavior, in turn, determines how (and if) rocks will move. We can classify the internal layers by physical properties as follows:

The *lithosphere* -- the Earth's cool, rigid outer layer -- may be up to about 100 - 200 kilometers (60 - 125 miles) in thickness. It is comprised of the brittle continental and oceanic crusts and the uppermost cool and rigid portion of the mantle.

The *asthenosphere* is the thin, hot, slowly-flowing layer of upper mantle below the lithosphere. Extending to a depth of about 350 - 650 kilometers (220 - 400 miles) the asthenosphere is characterized by its ability to deform under stress.

The *lower mantle* is the rigid middle and lower mantle extending to the core. Though it is hotter than the asthenosphere, the greater pressure at this depth probably prevents it from flowing.

The *core* is divided into two parts: the outer core is a viscous liquid with a density about 4 times that of the crust, the inner core a solid with a maximum density of about 6 times crustal material.

As we saw in the Chapter, recent research has shown that slabs of Earth's relatively cool and solid surface -- its lithosphere -- float and move independently of one another over the hotter, partially molten asthenosphere layer directly below. The physical properties of each make this possible, so classification by physical properties is more useful in explaining plate tectonics.

2. How is crust different from lithosphere?

Lithosphere includes crust (oceanic and continental) and rigid upper mantle down to the asthenosphere. The velocity of seismic waves in the crust is much different from that in the mantle. This suggests differences in chemical composition, or crystal structure, or both. The lithosphere and asthenosphere have different physical characteristics: the lithosphere is generally rigid, but the asthenosphere is capable of slow plastic movement. Asthenosphere and lithosphere also transmit seismic waves at different speeds.

3. Would the most violent earthquakes be associated with spreading centers or with subduction zones? Why?

Both sites are associated with earthquakes. Earthquakes at spreading centers (divergent boundaries) tend to be smaller and much more numerous than those experienced at subduction zones (convergent boundaries).

To use the Pacific as an example, consider the relatively calm spreading occurring at the East Pacific Rise. Though the divergence reaches 18 centimeters (7 inches) a year in places, the motion is accomplished with a minimum of fuss -- the warm seabed forms and moves outward with surprisingly little jerkiness.

At the other side of the Pacific, the cold, heavy, old, sediment-laden plate reaches the trench into which it will subduct. Pushed from behind (and pulled down by the weight of the subducting slab just ahead), the lithospheric plate will resist movement until motive forces overwhelm friction. Then, the trench south of the Aleutian Islands (or those southeast of the Kamchatka peninsula, or Japan) will swallow the leading edge of the plate in convulsive gulps, resulting in large and destructive earthquakes. Also, don't forget the action of volcanoes at convergent boundaries.

The Earth isn't growing, so the rate of convergence must equal the rate of divergence. But divergence is often a nearly continuous process, while convergence can be characterized by decades of calm punctuated by minutes of extreme geological excitement.

4. Describe the mechanism that powers the movement of the lithospheric plates.

The interior of Earth is hot; the main source of this heat is the radioactive decay of unstable elements within Earth.

When heated from below, the fluid asthenosphere expands, becomes less dense, and rises. It turns aside when it reaches the lithosphere, and drags the plates laterally until turning under again to complete the circuit. The large plates include both continental and oceanic crust. (The plates, which jostle about like huge flats of ice on a warming lake, are shown and named in Figure 3.15.) Plate movement is slow in human terms, averaging about 5 centimeters (2 inches) a year. The plates interact at converging, diverging, or slipping junctions, sometimes forcing one another below the surface or wrinkling into mountains. Most of the million or so earthquakes and volcanic events each year occur along plate boundaries.

Through the great expanse of geologic time this slow movement re-makes the surface of the Earth, expands and splits continents, forms and destroys ocean basins. This process has progressed since the Earth's crust first solidified.

5. Why are the continents about 20 times older than the oldest ocean basins?

The light, ancient granitic continents ride high in the lithospheric plates, rafting on the moving asthenosphere below. In subduction, heavy basaltic ocean floor (and its overlying layer of sediment) plunges into the mantle at a subduction zone to be partially remelted, but the light granitic continents ride above, too light to subduct. The subducting plate may be very slightly more dense than the upper asthenosphere on which it rides, and so is pulled downward into the mantle by gravity. Because the ocean floor itself acts as a vast "conveyor belt" transporting accumulated sediment to subduction zones where the seafloor sinks into the asthenosphere, no marine sediments (or underlying crust) are of great age. The ocean floor is recycled; the continents just jostle above the fray.

6. How does the principle of density stratification influence the plate tectonic process?

Gravity sorted our planet's components by density, separating Earth into layers (see again Figure 2.6). Because each deeper layer is denser than the layer above, we say Earth is density stratified. Remember, density is an expression of the relative heaviness of a substance; it is defined as the mass per unit volume, usually expressed in grams per cubic centimeter (g/cm^3). Remember, the lighter brittle lithosphere is supported by a deformable, moveable, but usually heavier layer (asthenosphere) beneath. Movement in

23

the lower layer can influence conditions in the lithosphere. Subduction occurs when conditions cause the oceanic lithosphere to become more dense (cooling, compression), enabling it to plunge beneath into a trench or beneath a continent. The subducting seabed's now-greater density pulls it toward the core-mantle boundary.

Thinking Critically

1. Seawater is more dense than fresh water. A ship moving from the Atlantic into the Great Lakes goes from seawater to fresh water. Will the ship sink farther into the water during the passage, stay at the same level, or rise slightly.

Buoyancy is the ability of an object to float in a fluid by displacing a volume of that fluid equal in mass to the floating object's own mass. A steel ship floats because its shape displaces a volume of water equal in weight to its own weight plus the weight of its cargo. A given volume of seawater is more dense than an identical volume of fresh water. If a ship ventures into water less dense than seawater (as would be the case upon entering the Great Lakes), it would need to sink into the less dense fresh water to displace a volume equal to its weight.

2. Some earthquakes are linked to adjustments of isostatic equilibrium. How can this occur? Where would you be likely to experience such an earthquake?

Remember, great continents and mountain ranges are not supported by the *mechanical* strength of the materials within the Earth -- nothing on (or in) our planet is that strong. Any region of a continent that projects above sea level is supported in isostatic equilibrium, a situation similar to buoyancy. Just as a ship rises when cargo is offloaded, a continent will tend to rise as material is eroded from its surface. Earthquakes resulting from adjustments to erosion can suddenly lift mountains a meter or so, as Charles Darwin experienced during his visit to the Andes. The Himalayas, Rocky Mountains, or Alps -- all places where erosion is rapidly altering the balance between gravity and buoyancy -- would be likely candidates for these types of earthquakes.

3. Where are the youngest rocks in the seabed? The oldest? Why?

The youngest rocks -- indeed, rocks still being formed -- are at the spreading centers; places like the East-Pacific Rise and the Mid-Atlantic Ridge. The oldest rocks are found beneath the layers of sediment descending into subduction zones in the northwestern Pacific. For a preview of these ages and directions of movement, please look ahead to Figure 5.23.

The age differential is caused by the conveyor-belt-like movement of the seabed characteristic of the plate tectonics process. Rocks are found to be progressively older as the distance from a spreading center increases.

4. Why did geologists have such strong objections to Wegener's ideas when he proposed them in 1912.

First, his training – he was a meteorologist, not a geologist. He was also an outsider with no university post in the Earth sciences, a detriment to being taken seriously in the European scientific establishment. Next, his evidence, though remarkable, proposed no rational motive force for "continental drift." He thought some sort of centrifugal "force" was responsible for the motion, but was himself unsure of its nature. Perhaps most tellingly, there was no evidence of the "tracks" left behind moving continents as they plowed along the seafloor (a vision we know to be false). Still, those jigsaw-puzzle-like fits at the edges of the continental shelves gnawed at his detractors' thoughts.

5. What biological evidence supports plate tectonics theory?

Wegener was right in suggesting the *Glossopteris* flora were an important piece of evidence in the puzzle. Animal fossils also support the idea of an ancient supercontinent. Fossils of *Mesosaurus*, a half-meter (2-foot) long aquatic reptile, are found only in eastern South America and southwestern Africa. It is extremely unlikely that this animal could have evolved simultaneously in two widely separated locations. It is equally unlikely that this small shallow-water reptile could have swum across 5,500 kilometers (2,500 miles) of open ocean to establish itself on both sides of the Atlantic.

Interestingly, present-day sea turtles also point to the operation of plate tectonics as described in the chapter. A population of green turtles (*Chelonia mydas*) lives off the coast of Brazil but regularly breeds 2,000 kilometers (1,235 miles) away on tiny Ascension Island, a projection of the Mid-Atlantic ridge. How could these animals make such a long journey to such a small target?

Perhaps the green turtles' distant ancestors had a much easier task when the Atlantic was very small. As seafloor spreading widened the ocean, their original breeding islands sank below sea level, but island after volcanic island erupted in the turtles' path to take their places. Successive generations would need only to extend their travel path directly into the rising sun to accommodate the growing ocean. As the distances grew, turtles adept at homing would have been favorably selected by the environment and would have reproduced most successfully. No other theory explains so well how the turtles' navigational accuracy evolved.

6. Imagine a tectonic plate moving westward. What geological effects would you expect to see on its northern edge? Western edge? Eastern edge?

Figure 3.16 covers this idea, but before turning to it, use your understanding of plate tectonics to predict an answer. The western edge would be a converging margin, and either mountains or a trench would form there. (Remember, the densities of the colliding plates determines which one, if either, will plunge beneath the other.) The eastern edge would be a divergent margin, and a new ocean will form there (if the convergence is on land), or an existing ocean will grow in size (at a spreading center). The northern edge is a transform (transverse) margin, a place where plates slip past each other. California's infamous San Andreas Fault is a transform margin.

7. What evidence can you cite to support the theory of plate tectonics? What questions remain unanswered? Which side would you take in a debate?

The evidence for plate tectonics includes the distribution and age of mid-ocean ridges, hot spots, and trenches; the configuration and location of atolls and guyots; the age of sediments; the presence of terranes at the edges of continental masses; fossils; and, of course, paleomagnetic data.

Questions remaining to be answered include: Why long lines of asthenosphere should be any warmer than adjacent areas; why the plastic material should flow parallel to the plate bottoms for long distances instead of cooling and sinking near the spreading center; whether plate movements are due entirely to motion of the asthenosphere, or the gravity-powered pulling of the descending plate is the major force; if the mantle circulates as a unit, or if it is also segregated into layers; if spreading always been a feature of the Earth's surface; and whether a previously thin crust become thicker with time, permitting plates to function in the ways described in this chapter.

If I had a choice, I would take the side of the "drifters."

Thinking Analytically

1. How much farther would Columbus have to sail if he crossed the Atlantic today?

I am writing this in 2004, so the elapsed time is 512 years. If we assume the rate of expansion of the Atlantic basin *on each side* of the Mid-

Atlantic Ridge has remained relatively constant at its present rate of about 7 centimeters (~3 inches) per year, the one-way trip would increase by about 14 centimeters (~6 inches) each year. Multiplying by the number of years (512) yields a spread of 7,168 centimeters or ~72 meters. At 236 feet, this is significantly less than the length of an American football field. I doubt Columbus would have noticed an extension of his travel time, even if he *had* known where he truly was.

2. Look at Figure 3.28. Knowing the Atlantic's spreading rate, how wide would the orange reversed-polarity blocks be in the figure?

It took 3 million years of continuous reversed polarity to form the orange blocks. If we use the data from the previous question (7 centimeters per year), we would calculate a width of (7 centimeters/year) x (3 million years) = 21 million centimeters = 210,000 meters = 210 kilometers (or 132 miles). The colored segments south and west of Iceland in Figure 3.26 are in scale.

CHAPTER 4

CONTINENTAL MARGINS AND OCEAN BASINS

Reviewing What You've Learned

1. How can satellites be used to measure ocean surface height? Why does the surface of the ocean "bunch up" over submerged mountains and ridges?

Satellites cannot measure ocean depths directly, but they can measure small variations in the elevation of surface water. Using about 1,000 radar pulses each second, satellites can measure its distance from the ocean surface to within 0.03 meters (1 inch)! Because the precise position of the satellite can be calculated, the average height of the ocean surface can be known with great accuracy.

Disregarding waves or tides or currents, the ocean surface can vary from the ideal smooth (ellipsoid) shape by as much as 200 meters (660 feet). This is because the pull of gravity varies across the surface of the Earth depending on the nearness (or distance away) of massive parts of the Earth. The mountain or ridge "pulls" water toward it from the sides, forming a mount of water over itself. For example, a typical undersea volcano with a height of 2000 meters (6600 feet) above the seabed, and a radius of 20 kilometers (32 miles), would produce a 2 meter (6.6 foot) rise in the ocean surface.

2. How does a multi-beam sonar system work?

Like an echo sounder, a multi-beam system bounces sound off the seafloor to measure ocean depth. Unlike a simple echo sounder, a multi-beam system may have as many as 121 beams radiating from a ship's hull. Fanning out at right angles to the direction of travel, these beams can cover a 120° arc. Typically, a pulse of sound energy is sent toward the seabed every 10 seconds. Listening devices record sounds reflected from the bottom, but only from the narrow corridors corresponding to the outgoing pulse. Successive observations build a continuous swath of coverage beneath the ship. By "mowing the lawn" – moving the ship in a coverage pattern similar to cutting grass – researchers can build a complete map of an area.

3. Imagine you could walk from the shore to the deep sea bed off the east coast of the United States. What features would you encounter?

You would begin you journey on the gently sloping continental shelf. You would not notice the gradual slope – typically 1.7 meters per kilometer (0.1°, or about 9 feet per mile), much less than the slope of a well-drained parking lot. About 350 kilometers from shore (220 miles) at a depth of 140 meters (460 feet), transition to the steeper continental slope begins. This location is known as the "shelf break." Depending on your position, you might see a submarine canyon slicing into the edge. Though noticeably steeper than the continental shelf, the continental slope is not unusually steep – typically about 4° (or 70 meters per kilometer, 370 feet per mile). An apron of accumulated sediment, the continental rise, would greet you at the bottom of the slope. Sometimes rough, sometimes smooth, you would need to climb across it to reach the flattest, most featureless feature on Earth: the abyssal plain. These expanses of sediment-covered seabed are found on the periphery of all ocean basins. Continued journeying would take you to the center of the Atlantic basin and the active mid-ocean ridge, where, with luck, you would encounter evidence of seafloor spreading and geothermal activity.

4. *What happened to continental shelves during ice ages?*

Because of their gentle slope, continental shelves are greatly influenced by changes in sea level. Around 18,000 years ago—at the height of the last ice age (periods of extensive glaciation)— massive ice caps covered huge regions of the continent. The water that formed the thick ice sheets was derived from the ocean, and sea level fell about 125 meters (410 feet) below its present position. The continental shelves were almost completely exposed, and the surface area of the continents was about 18% greater than it is today. Rivers and waves cut into the sediments that had accumulated during periods of higher sea level, and they transported some coarse sediments to their present locations at the shelves' outer edges. Sea level began to rise again when the ice caps melted, and sediments again began to accumulate on the shelves. (More on the history and effects of sea level change will be found in the discussion of coasts in Chapter 12.)

5. *Which is greater: The height of Mt. Everest or the depth of the Mariana Trench?*

The depth of the Mariana Trench, by about 20%.

6. *What do we think causes submarine canyons?*

Most geologists believe that the canyons have been formed by abrasive turbidity currents plunging down the canyons. Small amounts of debris may cascade continuously down the canyons, but earthquakes can shake loose huge masses of sediment that rush down the edge of the shelf, scouring the canyon deeper as they go. In this way the canyons can be cut to depths far below the reach of streams even during the low sea levels of the ice ages.

7. Why are abyssal plains relatively rare in the Pacific?

Because the extensive system of trenches along the active margins of the Pacific trap much of the sediments flowing off the continents, preventing them from building the broad, flat abyssal plains typical of the Atlantic. There are a few abyssal plains in the Pacific (notably adjacent to China and Southeast Asia), but none approaches the extent of, say, the Canary Abyssal Plain west of the Canary Islands in the North Atlantic, with an area of 900,000 square kilometers (350,000 square miles).

8. Why are trenches and island arcs curved? Is the descent to the bottom steeper on the convex side of the arc or on the concave side? Why the difference? Why do you think most trenches are in the western Pacific? (Hint: check the position and action of the East Pacific Rise.)

The trenches are curved because of the geometry of plate interactions on a sphere. To see a result of the deformation of a sphere, press on a ping pong ball with your thumb. The resulting depression will have curved edges; the convex sides of the curves will face your thumb. Note the similarity of that pattern to the position of the trenches in the northern and western Pacific (see Figure 4.29).

The convex sides of these curves face the open ocean. The trench walls on the island side of the depressions are steeper than those on the seaward side, indicating the direction of plate subduction. The sides of trenches become steeper with depth, normally reaching angles of about 10°-16° before flattening to a floor underlain by thick sediment. (Parts of the concave wall of the Kermadec-Tonga Trench are the world's steepest at 45°.) No continental rise occurs along coasts with trenches because the sediment which would form the rise ends up at the bottom of the trench.

Most trenches are in the subducting western Pacific opposite the East Pacific Rise, the spreading center responsible for the formation of most new Pacific seabed.

9. Distinguish among abyssal hills, seamounts, guyots, and island arcs.

Abyssal plains are often punctuated by abyssal hills, small sediment-covered extinct volcanoes or intrusions of once-molten rock less usually than 200 meters (650 feet) high. These abundant features are associated with sea floor spreading, and form when newly formed crust moves away from the center of a ridge, stretches, and cracks. Some blocks of the crust drop to

30

form valleys, and others remain higher as hills. Lava erupting from the ridge flows along the fractures, coating the hills. This helps explain why abyssal hills occur in lines parallel to the flanks of the nearby oceanic ridge, and occur most abundantly in places where the rate of seafloor spreading is fastest. Abyssal plains and abyssal hills account for nearly all of the area of deep ocean floor which is not part of the oceanic ridge system.

The ocean floor is dotted with thousands of "islands" that do not rise above the surface of the sea. These projections are called <u>seamounts</u>. Seamounts are circular or elliptical, usually more than a kilometer (0.6 mile) in height, with relatively steep slopes of 20° to 25°. (Abyssal hills, in contrast, are much more abundant, almost always smaller, and not as steep.) Seamounts may be found alone or in groups of from 10 to 100. Though many form at hot spots, most are thought to be submerged inactive volcanoes that formed at spreading centers. Movement of the lithosphere away from spreading centers has carried them outward and downward to their present positions. As many as 10,000 seamounts are thought to occur in the Pacific, about half the world total.

<u>Guyots</u> are flat-topped seamounts that once were tall enough to approach or penetrate the sea surface. Generally they are confined to the west central Pacific. The flat top suggests that they were eroded by wave action when they were near sea level. Their plateau-like tops eventually sank too deep for wave erosion to continue wearing them down. Like the more abundant seamounts, most guyots were formed near spreading centers and transported outward and downward by sea floor spreading.

<u>Island arcs</u>, curving chains of volcanic islands and seamounts, are almost always found paralleling the concave edges of trenches. As you may remember from Chapter 3, trenches and island arcs are formed by tectonic and volcanic activity associated with subduction. The descending lithospheric plate contains some materials that melt as the plate sinks into the mantle and rise to the surface as magmas and lavas that form the chain of islands behind the trench. The Aleutian Islands, most Caribbean islands, and the Marianas Islands are island arcs.

10. What is an oceanic ridge? Are they always literally in mid-ocean? How are oceanic ridges and trenches related?

An oceanic ridge is a mountainous chain of young basaltic rock at the active spreading center of an ocean. Stretching 65,000 kilometers (40,000 miles), more than 1 ½ times the Earth's circumference, oceanic ridges girdle the globe like seams surrounding a softball. Although these features are often called mid-ocean ridges, less than 60% of their length actually exists along the centers of ocean basins.

As we saw in our discussion of plate tectonics, the rift zones associated with oceanic ridges are sources of new ocean floor where lithospheric plates *diverge*. Trenches are found where plates *converge* and subduct. Seabed is born at the ridge, and dies in the trench.

Thinking Critically

1. Why did people think an ocean was deepest at its center? What changed their minds?

People who walked into the ocean were aware that the farther they walked, the deeper the water became (and the wetter they got). They saw boats close to shore, but knew that large ships could not venture that close without the danger of running aground. People reasonably assumed the gradual slope of the nearshore seabed continued into the depths, reached some hypothetical deepest spot near the middle, and then became progressively shallower until the opposite shore was reached.

Sporadic deep sampling provided hints that this "bathtub" model was not always true. During an expedition to scout the northwest passage in 1818, Sir John Ross obtained a series of bottom samples, the deepest of which was from 1,919 meters (3,296 feet) near Greenland. The soundings for these samples show an irregular bottom depth in the North Atlantic. A few of his log entries for the voyage reflect his puzzlement at finding the deepest parts of the North Atlantic near its periphery rather than at its center. Unfortunately, Ross was unable to take enough of the soundings to discern the contour of the seabed.

Sampling techniques improved through the century. Using a sounding method perfected in the late 1840s by a U.S. Navy midshipman, American Commodore Matthew Maury used a long lightweight line and lead weight to discover the Mid-Atlantic Ridge. But the breakthrough came in the form of the echo sounder, first employed in 1925 aboard the German research vessel *Meteor*. Scientists of the *Meteor* expedition criss-crossed the south Atlantic for two years, bouncing sound waves off the ocean bottom, studying the depth and contour of the seafloor. The echo sounder revealed a varied and often extremely rugged bottom profile rather than the flat floor they had anticipated. The central ridges found near the middle of the Atlantic have counterparts in nearly all ocean bottoms.

2. What do the facts that (a) granite underlies the edges of continents, and (b) basalt underlies deep ocean basins, suggest? (Hint: consider thicknesses and densities.)

Continents and deep ocean basins have different origins and different compositions.

We know that the undersea edges of continents are made of relatively light granitic rock buried beneath layers of sediment, and that the deep ocean floor is heavier basalt (also covered with sediment). The theory of plate tectonics explains why granite and basalt are distributed in that way. Deep soundings, echo sounders, and on-site observations have all contributed to our present understanding of ocean floor shape and structure. An example is shown in Figure 4.8 on page 93. Notice the abrupt transition between the thick granitic rock of the continents and the relatively thin basalt of the deep sea floor. Nearshore ocean floors are similar to the adjacent continents because they share the same granitic basement. The transition to basalt marks the true edge of the continent and divides ocean floors into two major provinces.

3. The terms leading and trailing are also used to describe continental margins. How do you suppose these words relate to active and passive, or Atlantic-type and Pacific-type used in the text?

The South American Plate is shown moving to the left (west) in Figure 4.10. The *leading* edge of the Plate -- the western edge -- is colliding with the Pacific seabed at the Peru-Chile Trench. This is obviously an active place (as any resident of the earthquake- and volcano-laden Andes can attest), and generally typical of much of the Pacific rim.

However, the characteristics of the leisurely progress of the *trailing* edge of the Plate -- the eastern edge -- away from the spreading center at the Mid-Atlantic Ridge is much calmer. This passive trailing edge, typical of nearly the whole Atlantic periphery, makes a geologically quiet contrast to the Pacific. Earthquakes and volcanic eruptions are big news in Tokyo, Seattle, and Mexico City, but one rarely reads of that kind of excitement in New York, Buenos Aires, or London.

4. What forces control the shape of a continental shelf? A continental slope? A continental rise?

As noted in the answer to the previous question, the forces associated with tectonic movement are certainly important in shaping continental shelves, slopes, and rises -- trailing margins have gentler and broader features. And, as we'll see in the next chapter, the deposition of sediments influences the contours greatly. Sediments result mainly from the accumulation of small hard parts of drifting marine organisms or from particles washed (or blown) from land. Ocean areas experiencing high biological productivity and erosive runoff from land tend to have high continental rises and flat, featureless continental shelves. Continental shelves can be slashed by submarine canyons formed (in part) due to the sliding of sediments off the edges of the shelves. Isostatic depression or rebound can also influence the position and contour of margin features -- remember the weight of ice atop Antarctica has pushed the continent (and its margin) down, and its shelf edges are among the deepest in the world.

5. Answer this question if you have already read Chapter 3: Your time machine has been programmed to deliver you to Frankfurt on a chilly evening in January 1912, to hear Wegener's lectures on continental drift. What two illustrations from this chapter would you take with you to cheer him up after the lecture? Why did you select those particular illustrations?

Dr. Alfred Wegener's presentation to the Geological Association of Frankfurt am Main on 6 January 1912 must have made an interesting evening. The tall, vigorous explorer-geologist was a forceful speaker, and his Frankfurt lecture was the first public announcement of his theory of "continental displacement." He began by noting, "...on studying the map of the world, I was impressed by the congruency of both sides of the Atlantic coasts, but I disregarded it at the time because I did not consider it probable." His theory posited the breakup of Pangaea (from Latin, "all Earth"), with its pieces -- our present continents -- plowing slowly into a single world ocean, Panthalassa (also from Latin, "all ocean"). His lecture that cold night was not particularly well received because he proposed no reasonable motive force for the drifting continents: "The question as to what forces have caused these displacements ... cannot yet be answered conclusively.   I can imagine he was disappointed in his theory's reception. He'd need some good cheer and, perhaps, an invitation to a warm restaurant for some hot wurst, some boiled red potatoes, and a beer.

The first illustration I'd take in my backpack (to spread out on the table after the dinner dishes were cleared) would be Figure 3.9. Wegener thought the "fit" of Atlantic continental edges was good at the shorelines, but continental shelf soundings had not been made in adequate number or accuracy to influence his work (until later). But look at the fit at the edges of the continental shelves! He would have loved it.

The second illustration would be Heinrich Berann's beautiful painting of the Atlantic Ocean Floor (Figure 4.22). I have no doubt that, at first, Wegener would have been horrified at the sight of it. "We do not know ... a single feature ... in the deep sea which we could claim with any certainty as a chain of mountains," he wrote. Such mountains would get in the way of the shoving continents, and their presence does great damage to his theory.

But Wegener was an imaginative and forward-looking scientist. He was not always right, of course, but he was always curious. Think of what fun you would then have telling him of the way we now think the Earth works: all the things you have learned so far in your oceanography course -- the differences between the continents and seabed, the great cycle of ocean beds opening and closing, the subduction zones and spreading centers, the Earth's vast age and origin, the ingenious devices used by modern scientists to discover new things about the ocean... If I were permitted to bring one more illustration, I'd leave him a copy of The Grand Tour, Figure 4.31 on page 108. Its presentation would top a truly memorable evening. Take me along – I have more than enough frequent-flyer miles!

Thinking Analytically

1. The speed of sound through seawater is about 1,500 meters per second. If a ship equipped with a multibeam mapping system is surveying a feature 3,500 meters below the surface, and if the researches wish to obtain an image of the feature at a resolution of 10 meters, what is the maximum speed the ship can steam?

The round-trip distance traversed by the "ping" (sonar pulse) would be 3,500 meters x (2), or 7,000 meters. At a speed of 1,500 meters/second, the ping's round-trip would take about 4.7 seconds. For the called-for resolution, the ship must not move more than 10 meters in 4.7 seconds, or 2.12 meters per second = 7.63 kilometers per hour (4.7 miles per hour).

2. Review the speed of a turbidity current. In the unlikely event that a fast-running current formed near the shoreline of a trailing edge coast, how long would it take for the current to traverse a typical continental shelf and arrive at the shelf break? Would you expect the current to move at a constant speed during this traverse?

A turbidity current running down a continental slope typically travels at speeds up to 27 kilometers (17 miles) per hour. A continental shelf is nearly flat, so a current would be unlikely to be sustained for long distances – certainly its speed would vary with small variations in incline. For the sake of calculation, let's assume an average speed of about 20 kilometers (12.5 miles) per hour. A typical trailing shelf is about 350 kilometers (220 miles) wide, so if the current moved continuously, a traverse would take (350 kilometers) divided by (20 kilometers per hour) = 17.5 hours.

3. How much wider has Iceland become because of seafloor spreading since the last huge sequence of eruptions in the 1500s? [Hint: Use the data in Figure 4.21.]

Figure 4.21 suggests the part of the Mid-Atlantic Ridge inhabited by Icelanders is spreading at a rate of 2.6 centimeters (on each side of the rift) per year. Now look at Figure 4.23 for a look at the rift itself. The total spread (left arrow plus right arrow) would be 2.6 centimeters per year for (2004 − 1500) = ~500 years. (500 years) x (2.6 centimeters/year) = 1,300 centimeters = 13 meters. How wide does that rift in Figure 4.23 look (note the road). When do you think that rift began to split?

Chapter 5

Sediments

Reviewing What You've Learned

1. In what ways are sediments classified?

Sediments are generally classified by <u>particle size</u> and by <u>origin</u>.

In classification by particle size, the coarsest particles are boulders, which are more than 256 millimeters (about 10 inches) in diameter. Although boulders, cobbles, and pebbles occur in the ocean, most marine sediments are made of finer particles: sand, silt, and clay. The smaller the particle, the more easily it can be transported by streams, waves, and currents. As sediment is transported it tends to be sorted by size -- coarser grains, which are only moved by turbulent flow, tend to remain behind finer grains, which are more readily moved. The clays, particles less than 0.004 millimeters in diameter, can remain suspended for very long periods and may be transported great distances by ocean currents before they are deposited.

Another way to classify marine sediments is by their origin. Such a scheme was first proposed in 1891 by Sir John Murray and A. F. Renard after a thorough study of sediments collected during the *Challenger* expedition. A modern modification of their organization by origin is shown in Table 5.2. This scheme separates sediments into four categories by source: terrigenous, biogenous, hydrogenous (or authigenic), and cosmogenous. These categories are discussed in the answer to the next question.

2. List the four types of marine sediments. Explain the origin of each.

Marine sediments are separated into four categories by source: terrigenous, biogenous, hydrogenous (or authigenic), and cosmogenous. <u>Terrigenous sediments</u> are the most abundant. As the name implies, terrigenous sediment originates on the continents or islands near them. They are carried to the ocean in rivers and streams, or by winds as blowing dust, and dominate the continental margins, abyssal plains, and polar ocean floors. <u>Biogenous sediments</u>, the next most abundant, consist of the hard remains of once-living marine organisms. The siliceous (silicon-containing) and calcareous (calcium carbonate-containing) compounds that make up these sediments of biological origin were originally dissolved in the ocean at mid-ocean ridges or brought to the ocean in solution by rivers. Biogenous sediments are found mixed with terrigenous material near continental margins, but are dominant on the deep ocean floor. <u>Hydrogenous sediments</u> are minerals that have precipitated directly from seawater. The sources of the dissolved minerals include submerged rock and sediment, leaching of the fresh crust at oceanic ridges, material issuing from hydrothermal vents, or substances flowing to the ocean in river runoff. The most prominent hydrogenous sediments are manganese nodules, which litter abyssal plains, and phosphorite nodules, seen along some continental margins. Hydrogenous sediments are also called authigenic because they were formed in the place they now occupy. <u>Cosmogenous sediments</u>, which are of extraterrestrial origin, are the least abundant. These particles enter the Earth's high atmosphere as blazing meteors or as quiet motes of dust. Their rate of accumulation is so slow that they never accumulate as distinct layers -- they occur as isolated grains in other sediments, rarely constituting more than 1% of any layer.

3. How are neritic sediments generally different from pelagic ones?

Remember that sediments on the ocean floor only rarely come from a single source; most sediment deposits are a mixture of particles. The patterns and composition of sediment layers on the seabed are of great interest to researchers studying conditions in the overlying ocean. Different marine environments have characteristic sediments, and these sediments preserve a record of past and present conditions within those environments.

The sediments on the continental margin are generally different in quantity, character, and composition from those on the deeper basin floors. Continental shelf sediments -- neritic sediments -- consist primarily of terrigenous material. Deep ocean floors are covered by finer sediments than those of the continental margins, and a greater proportion of deep sea sediment is of biogenous origin. Sediments of the slope, rise, and deep ocean floor that originate in the ocean are called pelagic sediments. The distribution and average thickness of the marine sediments in each oceanic region are shown in Table 5.3 in your textbook.

4. Where are sediments thickest? Are there any areas of the ocean floor free of sediments?

Lithified sediments can be miles thick. As you may recall, much of the Colorado Plateau with its many stacked layers was formed by sedimentary deposition and lithification beneath a shallow continental sea beginning about 570 million years ago. The Colorado River has cut and exposed the uplifted beds to form the Grand Canyon. Hikers walking from the Canyon rim down to the river pass through spectacular examples of continental shelf sedimentary deposits. Most of the upper sediments have already been eroded, but the remaining material is more than 1 mile (1.6 kilometers) deep.

The loose sediments of the Continental Rise (at the foot of the Continental Shelf), transported into position by turbidity currents, may reach depths of 10 kilometers (6.2 miles).

No sediments can accumulate in areas where swift deep currents scour the seabed, and the fresh rock of the mid-ocean ridges -- in the rifts of spreading centers -- is free of sediments for a short time after its formation.

5. Are sediments commercially important? In what ways?

Study of sediments has brought practical benefits. In 1998 32% of the world's crude oil and 24% of its natural gas were extracted from the sedimentary deposits of continental shelves and continental rises. Offshore hydrocarbons presently generate annual revenues in excess of $125 billion. Deposits within the sediments of continental margins account for nearly one-third of the world's estimated oil and gas reserves.

In addition to oil and gas, in 2000 (the last year for which I have good data) sand and gravel valued at more than $510 million was taken from the ocean. This is about 1% of world needs. Commercial mining of manganese nodules has been considered. In addition to manganese, these chunks also contain substantial amounts of iron and other industrially important chemical elements. The high iron content of these nodules has prompted a proposal to re-name them ferromanganese nodules.

Thinking Critically

1. Is the thickness of ooze always an accurate indication of the biological productivity of surface water in a given area? (Hint: see next question.)

Not always. Currents may carry the residue from productive waters a great distance before permitting deposition. Rapidly moving bottom currents might scour sediments from some areas (especially on the shelf edges and slopes beneath western boundary currents). And some of the fragments of biological material might dissolve before reaching the bottom (as discussed in the next question).

2. What is the calcium carbonate compensation depth? Is there a compensation depth for the siliceous components of once-living things?

Calcareous ooze forms mainly from shells of the amoeba-like foraminifera, small drifting mollusks called pteropods, and tiny algae known as coccolithophores. Though these creatures live in nearly all surface ocean water, calcareous ooze does not accumulate everywhere on the ocean floor because the shells are dissolved by seawater. At great depths seawater contains more CO_2 and becomes slightly acid. This acidity, combined with the increased solubility of calcium carbonate in cold water under pressure, dissolves the shells as they fall. At a certain depth, the <u>calcium carbonate compensation depth</u>, the rate at which calcareous sediments accumulate equals the rate at which those sediments dissolve. Below this depth the tiny skeletons of calcium carbonate dissolve as they fall (or soon after they reach the bottom) so no calcareous oozes form. Calcareous sediment dominates at bottom depths less than about 4,500 meters (14,800 feet), the usual calcium carbonate compensation depth. Sometimes a "snow line" can be seen on undersea peaks -- above the line the white sprinkling of calcareous ooze is visible; below it, the "snow" is absent (see Figure 5.14). About 48% of all deep ocean sediments are calcareous oozes.

Siliceous (silicon-containing) ooze predominates at greater depths and in colder polar regions. Siliceous ooze is formed from the hard parts of another amoeba-like animal, the beautiful glassy radiolarian, and from single-celled algae, diatoms. The silica within these organisms can dissolve in deep water (which is usually deficient in silicon), but silica dissolves *much* more slowly than calcium carbonate does -- there is no silica compensation depth. Slow dissolution, combined with very high diatom productivity in surface waters, leads to the buildup of siliceous ooze.

3. What sediments accumulate most rapidly? The least rapidly?

Accumulation rate depends on the availability of the sediment in question. The rate of sediment deposition on continental shelves is variable, but it is almost always greater than the rate of sediment deposition in the deep ocean. Near the mouths of large rivers, 1 meter (about 3 feet) of terrigenous sediment may accumulate every 1,000 years. In the deep ocean, mudslides rushing down the continental slope deposit turbidites -- layers of coarse-grained terrigenous sediments interleaved with finer sediments typical of the deep-sea floor. Turbidite accumulation may be quite rapid adjacent to continental shelves shaken by earthquakes and subject to much erosional runoff from land.

The sediments slowest to accumulate are hydrogenous sediments. Accumulation rates on manganese nodules are typically the thickness of a dime every thousand years. (The rate of accumulation of cosmogenous sediment is so slow that they never accumulate as distinct layers. They occur as isolated grains in other sediments, rarely constituting more than 1% of any layer.)

4. Can marine sediments tell us about the history of the ocean from the time of its origin? Why or why not?

The distribution, depth, and composition of sediment layers tell of conditions in the comparatively recent past. Figure 5.23 in your textbook shows the age of the Pacific ocean floor using data obtained largely from analyses of the overlying sediment. Note that sediments get older with increasing distance from the East Pacific Rise spreading center, but the maximum age is roughly early Cretaceous or late Jurassic (around 145 million years old). The "memory" of the sediments is not ancient and in fact is continually being erased by ocean floor subduction. We can't see farther back than about 180 million years because the oceanic conveyor belt of plate tectonic processes destroys the evidence.

Still, marine sediments in the modern basins can shed light on unexpected details of the last 180 million years of Earth's history. One of the oddest details is the unexplained extinction of up to 52% of known marine animal species (and the dinosaurs) at the end of the Cretaceous Period 65 million years ago. Researchers have proposed hypotheses such as a sudden and violent increase in worldwide volcanism or the impact of one or more very large meteors or comets to explain this catastrophe. The clouds of dust and ash thrown into the atmosphere by any of these events would have drastically reduced incident sunlight and greatly affected the lives of organisms and the photosynthetic base of ecosystems. Oceanographers are presently searching for evidence of the cause of the Cretaceous extinctions in layers of deep sediments.

5. What problems might arise when working with deep-ocean cores? Imagine the process of taking a core sample, and think of what can go wrong!

Murphy's Law is rarely more functional than in deep-water sampling. Where, exactly, is the end of the drill string? Will the sample be washed out of the coring tube as the array is lifted to the surface? Will the drill string snap and plunge to the bottom never to be seen again? Will we strike a hard rock instead of soft muck and bend the tip? Will ocean conditions in the water column we're passing through to the surface chemically affect the substances in the sample? You could go on and on...

Thinking Analytically

1. Given an average rate of accumulation, how much time would it take to build an average-size manganese nodule?

These are among the slowest chemical reactions in nature. If a potato-size nodule (say 12 centimeters in diameter) accumulates at a relatively high rate – 10 millimeters (= 1 centimeter) per million years – it would take 12 million years to form. An equally intriguing question is, "If they grow so slowly, why aren't they covered by terrigenous or biogenous sediment?" The main reason seems to be bioturbation – the churning of sediments by burrowing organisms just below the surface. When you shake a bottle filled with different sized marbles, the large ones come to the top. The same thing appears to happen on the sea floor.

2. Assuming an average rate of sediment accumulation and seafloor spreading, how far from the Mid-Atlantic Ridge would one need to travel before encountering a layer of sediment 1,000 meters thick?

The average thickness (ocean-wide) of the Atlantic seabed is about 1,000 meters (= 1 kilometer), but local variations in the rates of sedimentation accumulation are quite large. If we assume (and it's an "iffy" assumption) that an Atlantic deep-basin average accumulation would be in the ballpark of 0.01 centimeter (0.1 millimeter) per year, and if we further assume a spreading rate of 3 centimeters per year, then we expect to accumulate 0.1 millimeters of sediment for every 3 centimeters of travel. To extend this step by step, we'd see 1 millimeter in 30 horizontal centimeters, 1 centimeter in 300 horizontal centimeters (= 3 meters), 10 centimeters in 30 meters, 1 meter in 300 meters, and 1,000 meters in 300,000 meters' (= 3,000 kilometers') distance from the Mid-Atlantic Ridge.

3. Microtektites are often found in "fields"—elongated zones of relative concentration a few hundred kilometers long. Why do you suppose that is?

Throw a round rock hard into moist sand. Watch the spray pattern of the sand – it lands in a narrow fan pattern at a distance from the crater. Imagine an asteroid hitting Earth. The excavated crust (and, depending on impactor size, mantle) would spray away from the direction of impact in a similar pattern. Clumps of material would tend to stay together, and end up in elongated microtektite fields.

4. How much faster do fragments of diatoms fall when they are compacted into fecal pellets than when they are not? (Hint: See Table 5.1)

If we assume that individual diatom fragments are in the size-range of clay, and compacted fecal pellets are similar in size to silt (or, sometimes, small sand grains), the difference is 1:100. (That is, a ratio of 6 months to 50 years.)

CHAPTER 6

WATER AND OCEAN STRUCTURE

Reviewing What You've Learned

1. Why is water a polar molecule? What properties of water derive from its polar nature?

The angular shape of the water molecule makes it electrically asymmetrical, or polar. Each water molecule can be thought of as having a positive (+) end and a negative (-) end because positively charged particles at the center of the hydrogen atoms -- protons -- are left partially exposed when the negatively charged electrons bond more closely to oxygen. The polar water molecule acts something like a magnet; its positive end attracts particles having a negative charge, and its negative end attracts particles having a positive charge. When water comes into contact with compounds whose elements are held together by the attraction of opposite electrical charges (most salts, for example), the polar water molecule will separate that compound's component elements from each other. This explains why water can dissolve so many other compounds so easily.

The polar nature of water also permits it to attract other water molecules. When a hydrogen atom (positive end) in one water molecule is attracted to the oxygen atom (negative end) of an adjacent water molecule, a hydrogen bond forms. Hydrogen bonds greatly influence the properties of water by allowing individual water molecules to stick to each other, a property called cohesion. Cohesion gives water an unusually high surface tension, which results in a surface "skin" capable of supporting needles, razor blades, and even walking insects. It also causes the capillary action that makes water spread through a towel when one corner is dipped in water. Adhesion, the tendency of water to stick to other materials, allows water to adhere to solids, that is, to make them wet. The absorption of red light by hydrogen bonds is also what gives pure water -- and thick ice -- its pale bluish hue.

2. How is heat different from temperature?

Heat is energy produced by the random vibration of atoms or molecules. On the average, water molecules in hot water vibrate more rapidly than water molecules in cold water. Heat and temperature are not the same thing. Heat tells us *how many* and *how rapidly* molecules are vibrating. Temperature records only *how rapidly* the molecules of a substance are vibrating. Temperature is an object's response to an input (or removal) of heat. The amount of heat required to bring a substance to a certain temperature varies with the nature of that substance.

Do you recall the example given in the text? Which has a higher *temperature* -- a candle flame or a bathtub of hot water? The flame. Which contains more *heat*? The tub. The molecules in the flame vibrate very rapidly, but there are relatively few of them. The molecules of water in the tub vibrate more slowly, but there are a great many of them, so the total amount of heat energy in the tub is greater.

3. What is the difference between sensible and nonsensible heat? What is meant by latent heat?

A detectable decrease in heat -- that is, a decrease that can be measured by a thermometer or by your hand -- is called <u>sensible heat</u> loss. But the loss of heat as water freezes is not measurable (that is, <u>non-sensible</u>) by a thermometer. Removing a calorie of heat from freezing water at 0°C (32°F) won't change its temperature at all; 80 calories of heat energy must be removed per gram of pure water at 0°C (32°F) to form ice. This heat is called the <u>latent heat</u> of fusion (note that the Latin *latere* means to be hidden). The straight line between points C and D in Figure 6.6 freezer has turned to ice. If the removal of heat continues, the ice will get colder and will soon reach the temperature inside the freezer, point E in Figures 6.3 and 6.6.

4. Why does ice float? Why is this fact important to thermal conditions on Earth?

During the transition from liquid to solid state at the freezing point, the bond angle between the oxygen and hydrogen atoms in water expands from about 105° to slightly more than 109°. This change allows ice to form a crystal lattice (as seen in Figures 6.5 on page 142). The space taken by 24 water molecules in the solid lattice could be occupied by 27 water molecules in liquid state, so water expands about 9% as the crystal forms. Because the molecules are packed less efficiently, ice is less dense than liquid water and floats. Ice at 0°C (32°F) weighs only 0.917 g/c m^3 where liquid water at 0°C weighs 0.999 g/cm^3.

Because water expands and floats when it freezes, ice can absorb the morning warmth of the sun, melt, then re-freeze at night giving back to the atmosphere the heat it stored through the daylight hours. The *heat content* of the water changes through the day; its *temperature* does not. The same principle applies to the seasonal formation and melting of polar ice. More than 18,000 cubic kilometers (4,300 cubic miles) of polar ice thaws and refreezes each year. Seasonal extremes are moderated by the immense amounts of heat energy that are alternately absorbed and released without a change in temperature. Without these properties of ice, temperatures on the Earth's surface would change dramatically with minor changes in atmospheric transparency or solar output.

5. How is heat transported from tropical regions to polar regions?

Water's high heat capacity makes it an ideal fluid to equalize the polar-tropical heat imbalance. Ocean currents and atmospheric weather result from the response of water and air to unequal solar heating.

Ocean currents carry heat from the tropics (where incoming energy exceeds outgoing) to the poles (where outgoing energy exceeds incoming). The amount of heat transferred in this way is astonishing. As noted in the chapter, "outbound" water in the warm Gulf Stream (a large northward-flowing ocean current just offshore of the eastern United States) is about 10°C (18°F) warmer than "inbound" returning water, meaning that about 10 million calories are transported per cubic meter. Since the flow rate of the Gulf Stream is about 55 million cubic meters per second, some 550 trillion calories are being transported northward in the western North Atlantic each *second*! Nearly half of these calories reach the high latitudes above 40°N. As we will see in the opener for Chapter 9, this warmth has a dramatic moderating influence on the winter climate of northwestern Europe.

As impressive as the figures are for ocean currents, the amount of heat transported by water vapor in the atmosphere is even greater. About half of the solar energy entering water results in evaporation. The solar energy required for this evaporation is later surrendered during condensation, but usually at a distance from where the initial evaporation occurred. The ocean surface near Cuba may be cooled by evaporation today, the water vapor moved north by winds, and eastern Canada warmed by condensation of the same water in a rainstorm later in the week.

Both atmosphere and ocean transfer heat by movement, but water's exceptionally high latent heat of evaporation means that water vapor transfers much more heat (per unit of mass) than liquid water. Weather accounts for about two-thirds of the poleward transfer of heat; ocean currents move the other third.

6. What factors affect the density of water? Why does cold air or water tend to sink? What role does salinity play?

The density of water is mainly a function of its salinity and temperature. Cold, salty water is denser than warm, less salty water. The density of seawater varies between 1.020 and 1.030 g/cm^3, indicating that a liter of seawater weighs between 2% and 3% more than a liter of pure water (1.00

g/cm^3) at the same temperature. Seawater's density increases with increasing salinity, increasing pressure, and decreasing temperature.

Dense fluids (water or air) sink in the presence of less dense fluids because they weigh more per unit of volume than the surrounding fluid.

7. How is the ocean stratified by density? What physical factors are involved? What names are given to the ocean's density zones?

Ocean water tends to form into stable layers with the heaviest water at the bottom, a form of density stratification. The primary physical factors in determining density, as noted above, are temperature, salinity, and pressure.

Much of the ocean is divided into three density zones. The surface zone, or mixed layer, is the upper layer of ocean in which temperature and salinity are relatively constant with depth because of the action of waves and currents. The surface zone consists of water in contact with the atmosphere and exposed to sunlight; it contains the ocean's least dense water, and accounts for about 2% of total ocean volume. Depending on local conditions, the surface zone may reach a depth of 1,000 meters or be absent entirely. Beneath it is the pycnocline, a zone in which density increases with increasing depth. This zone isolates surface water from the more dense layer below. The pycnocline contains about 18% of all ocean water. The deep zone lies below the pycnocline at depths below about 1,000 meters (3,000 feet) in mid-latitudes (40°S to 40°N). There is little additional change in water density with increasing depth through this zone. This deep zone contains about 80% of all ocean water.

8. What factors influence the intensity and color of light in the sea? What factors affect the depth of the photic zone? Could there be a photocline in the ocean?

As noted above, sunlight has a difficult time reaching and penetrating the ocean -- clouds and the sea surface reflect light, atmospheric gases and particles scatter and absorb it. Once past the sea surface light is rapidly weakened by scattering and absorption. Scattering occurs as light is bounced between air or water molecules, dust particles, water droplets, or other objects before being absorbed. The absorption of light is governed by the structure of the water molecules it happens to strike. When light is absorbed, molecules vibrate and the light's energy is converted to heat.

Even perfectly clear seawater is not perfectly transparent. If it were, the sun's rays would illuminate the greatest depths of the ocean and seaweed forests would fill its warmed basins. The thin film of lighted water at the top of the surface zone is called the photic zone. In clear tropical waters the photic zone may extend to a depth of 200 meters (660 feet), but a more typical value for the open ocean is 100 meters (330 feet). Here water is heated by the sun, heat is transferred from the ocean into the atmosphere and space, and gases are exchanged with the atmosphere. The thermostatic effects we've discussed function largely within this zone. The ocean below the photic zone lies in blackness.

The light energy of some colors is converted into heat nearer to the surface than the light energy of other colors. Figure 6.20 shows this differential absorption by color. Notice that after 1 meter (3.3 feet) of travel, only 45% of the light energy remains, most of it in the green and blue wavelengths. After 10 meters (33 feet) 85% of the light has been absorbed, and after 100 meters (330 feet) just 1% remains. The dimming light becomes bluer with depth because the red, yellow, and orange wavelengths have already been absorbed. Even in the clearest conditions sunlight rarely penetrates below 250 meters (820 feet).

The "photocline," if such a thing could be considered to exist, would begin immediately at the ocean's surface and end at the bottom of the photic zone. Unlike the thermocline, there would be no equivalent of a mixed layer at the ocean surface.

9. Which moves faster through the ocean – light or sound?

Light travels <u>much</u> faster (but not as far) as sound. The speed of light in water is only about three-quarters its speed in air, about 223,000 kilometers per second (140,000 miles per second). The speed of sound in seawater is about 1,500 meters per second (3,345 miles per hour), almost five times the speed of sound in air.

10. How is the pycnocline related to the thermocline and halocline?

The pycnocline is usually the sum of the density variation with depth caused by temperature and salinity. That is, the pycnocline is the effective sum of the thermocline and halocline.

Thinking Critically

1. What is the latent heat of vaporization? Of fusion? Which requires more heat to transform water's physical state? Why?

When water vaporizes (or evaporates), individual water molecules diffuse into the air (Figure 6.6 and 6.7). Since each water molecule is hydrogen-bonded to adjacent molecules, heat energy is required to break those bonds and allow the molecule to fly away from the surface. Evaporation cools a moist surface because departing molecules of water vapor carry this energy away with them. Hydrogen bonds are quite strong, and the amount of energy required to break them—known as the latent heat of vaporization—is very high. At 540 calories per gram, water has the highest latent heat of vaporization of any known substance. As before, the term *latent* applies to heat input that does not cause a temperature change but does produce a change of state—in this case from liquid to gas.

Why the big difference between water's latent heat of *fusion* (80 calories per gram) and its latent heat of *vaporization* (540 calories per gram)? Only a small percentage of hydrogen bonds are broken when ice melts, but *all* must be broken during evaporation. Breaking these bonds requires additional energy in proportion to their number.

By the way, if the difference between latent heat of evaporation (585 calories per gram) and latent heat of vaporization (540 calories per gram) confuses you, see "Questions from Students" #2 on page 159 of your text.

2. How does water's high latent heat influence the ocean? Leaving aside its salubrious effect on beach parties, how do you think conditions on Earth would differ if our ocean consisted of ethyl alcohol?

The most important effect of water's high latent heat (or heat capacity) is thermal inertia, the tendency of a substance to resist change in temperature with the gain or loss of heat energy. Liquid water's high heat capacity (and therefore large thermal inertia) prevents broad swings of temperature during day and night, and, through a longer span, during winter and summer. Heat is stored in the ocean during the day and released at night. A much greater amount of heat is stored through the summer, and given off during the winter.

By contrast with water, ethyl alcohol has a much lower latent heat. If both liquids absorb heat from identical stove burners at the same rate, pure ethyl alcohol, the active ingredient in alcoholic beverages, will rise in temperature about *twice* as fast as an equal mass of water. If our ocean were made of alcohol -- or almost any other liquid -- summer temperatures would be much hotter, and winters bitterly cold. Storms would be much more violent because the thermal imbalance between the summer and winter hemispheres would be greater and winds would be stronger.

3. How does the seasonal freezing and thawing of the polar ocean areas influence global temperatures?

As you know, removing a calorie of heat from freezing pure water at 0°C (32°F) won't change its temperature at all -- 80 calories of heat energy must be removed per gram of liquid water to form ice. More than 18,000 cubic kilometers (4,300 cubic miles) of polar ice, covering as many as 20 million square kilometers (7.7 million square miles) of surface, thaws and refreezes in the Southern Hemisphere each year -- an area of ocean larger than South America (Figure 6.11)! The annual change in sea ice cover is less in the Arctic, averaging about 5 million square kilometers (2 million square miles). Imagine the vast amounts of energy that can be absorbed and released as the polar ocean areas thaw and re-freeze each summer and winter. Without this latent heat, temperatures over the world would be much more extreme – warmer summers and cooler winters. Global winds would be much stronger as thermal imbalances were distributed.

4. How does refraction permit sound to be transmitted in the ocean for thousands of miles?

The depth at which the speed of sound reaches its minimum varies with conditions, but is usually located near 1,200 meters (3,900 feet) in the North Atlantic or about 600 meters (2,000 feet) in the North Pacific. Transmission of sound in this minimum-velocity layer is very efficient because refraction tends to cause sound energy to remain within the layer. The outer edges of sound waves escaping from this layer will enter water in which the speed of sound is higher, speed up, and cause the wave to pivot back into the minimum-velocity layer, as shown in Figure 6.22). Upward-traveling sound waves that are generated within the minimum-velocity layer will tend to be refracted downward, and downward-traveling sound waves will tend to be refracted upward. In short, sound waves bend *toward* layers of lower sound velocity and so tend to stay within the zone. Therefore, loud noises made at this depth can be heard for thousands of kilometers. Indeed, Navy depth charges detonated in the minimum velocity layer in the Pacific have been heard 3,680 kilometers (2,280 miles) from the explosion.

5. What factors influence the density of seawater?

The density of water is mainly a function of its salinity and temperature. Cold, salty water is denser than warm, less salty water. The density of seawater varies between 1.020 and 1.030 g/cm^3, indicating that a liter of seawater weighs between 2% and 3% more than a liter of pure water (1.00 g/cm^3) at the same temperature. Seawater's density increases with increasing salinity, increasing pressure, and decreasing temperature.

Thinking Analytically

1. How many Calories (note capital "C") are required to raise the temperature of a can of diet cola (12 fluid ounces) from refrigerator temperature to body temperature. What percentage of the energy (Calories) in the can of diet cola is required to do this? Assume, for the moment, that the diet cola has the same thermal properties as pure water. Hint: See footnote 1 on page 139.

A nutritional Calorie (C) is 1,000 calories. Let's do the math in calories, and convert at the end.

Twelve fluid ounces (U.S.) = 29.6 cubic centimeters, so the can contains 355 cubic centimeters of soda. Refrigerator temperature is about 35°F (1.7°C), and body core temperature is about 98°F (37°C), so the temperature rise would be about 35°C. A temperature rise of 35°C for one cubic centimeter would require a caloric input of 35 calories, so 355 cubic centimeters would require 35 x 355 = 12,425 calories. Since 1 Calorie = 1,000 calories, 12.4 nutritional Calories would be required. My diet soda can says "0 Calories," so I would expend calories (and thus lose weight) to heat the fluid to my temperature.

What about regular cola? A can of Classic Coke contains 140 Calories (140,000 calories). If we assume sugared cola has the same properties as pure water (it doesn't), we would use only about 9% of the energy (12.4 is 9% of 140) in the non-diet soda to warm the fluid to core temperature – the rest would be used as energy (or stored as fat).

2. How much heat is required to melt a quart of ice resting at 0°C?

A quart of water is 946 cubic centimeters or 946 grams. The latent heat of fusion of pure water is 80 calories per gram. 946 × 80 = 75,680 calories.

3. How much heat is required to vaporize a quart of pure water that has just come to a boil?

A quart of water is 946 cubic centimeters or 946 grams. The latent heat of vaporization is 540 calories per gram. 946 × 540 = 510,840 calories.

CHAPTER 7

SEAWATER CHEMISTRY

Reviewing What You've Learned

1. Why is water a polar molecule? What properties of water derive from its polar nature?

The angular shape of the water molecule makes it electrically asymmetrical, or polar. Each water molecule can be thought of as having a positive (+) end and a negative (-) end because positively charged particles at the center of the hydrogen atoms -- protons -- are left partially exposed when the negatively charged electrons bond more closely to oxygen. The polar water molecule acts something like a magnet; its positive end attracts particles having a negative charge, and its negative end attracts particles having a positive charge. When water comes into contact with compounds whose elements are held together by the attraction of opposite electrical charges (most salts, for example), the polar water molecule will separate that compound's component elements from each other. This explains why water can dissolve so many other compounds so easily.

The polar nature of water also permits it to attract other water molecules. When a hydrogen atom (positive end) in one water molecule is attracted to the oxygen atom (negative end) of an adjacent water molecule, a hydrogen bond forms. Hydrogen bonds greatly influence the properties of water by allowing individual water molecules to stick to each other, a property called cohesion. Cohesion gives water an unusually high surface tension, which results in a surface "skin" capable of supporting needles, razor blades, and even walking insects. It also causes the capillary action that makes water spread through a towel when one corner is dipped in water. Adhesion, the tendency of water to stick to other materials, allows water to adhere to solids, that is, to make them wet. The absorption of red light by hydrogen bonds is also what gives pure water -- and thick ice -- its pale bluish hue.

2. How does a crystal of common salt dissolve in water?

Unlike the electron *sharing* found in covalently bonded molecules, the sodium atoms in NaCl have *lost* electrons, and chlorine atoms have *gained* them. The resulting ions are linked by the mutual attraction of their opposite electrical charges. The ions of sodium and chloride in NaCl are said to be held together by ionic bonds, electrostatic attraction that exists between ions that have opposite charge. When NaCl dissolves in water (Figure 7.3), the polarity of water reduces the electrostatic attraction (ionic bonding) between the sodium ion (Na^+) and chloride ion (Cl^-). This causes the sodium ion to separate from the chloride ion. The ions move away from the salt crystal, permitting water to attack the next layer of NaCl.

Note that NaCl does not exist as "salt" in seawater; its components are separated when salt crystals dissolve in water, but they are joined when crystals re-form as water evaporates.

3. How does a solution differ from a mixture?

A <u>solution</u> is made of two components: the solvent, usually a liquid, is always the more abundant constituent; the solute, often a dissolved solid or gas, is the less abundant. In a true solution (sugar in well-stirred coffee, for example), the molecules of the solute are homogeneously dispersed among the molecules of solvent; that is, the solution has uniform properties throughout. In a <u>mixture</u> different substances are closely intermingled but retain separate identities. The properties of a mixture are heterogeneous; they may vary from place to place within the mixture. Think of noodle soup as a mixture.

4. What are some of seawater's colligative properties? Does pure water have colligative properties?

The many ions present in seawater react with each other (and with water molecules) in complex ways to modify the physical properties of pure water. Properties which vary with the quantity of solutes dissolved in the water are called water's <u>colligative properties</u>. Because colligative properties are the properties of *solutions*, the more concentrated (saline) water is, the more important these properties become. (Because it is not a solution, pure water has no colligative properties.)

Examples of water's colligative properties are: (1) The heat capacity of water; (2) depression of seawater's freezing point; (3) slower evaporation of seawater; (4) osmotic pressure, the pressure exerted on a biological membrane when the salinity of the environment is different from that within the cells.

5. Other than hydrogen and oxygen, what are the most abundant elements in seawater?

In order, the five most abundant <u>elements</u> in seawater are sodium, chlorine, maganesium, sulfur, and calcium. More useful to oceanographers, however, is the abundance of <u>ions</u>, atoms or groups of atoms that become electrically charged by gaining or losing one or more electrons. Seven ions make up more than 99% of the non-water material of seawater. They are listed here in order of their concentrations. (For more detail, please see Table 7.1 and Figure 7.4.)

Chloride (Cl^-)

Sodium (Na^+)

Sulfate (SO_4^{2-})

Magnesium (Mg^{2+})

Calcium (Ca^{2+})

Potassium (K^+)

Bicarbonate (HCO_3^-)

6. How is salinity determined?

Modern analysis of salinity depends on determining a seawater sample's chlorinity, or measuring its electrical conductivity.

Chlorinity is a measure of the total weight of chlorine, bromine, and iodine ions in seawater. Because chlorinity is comparatively easy to measure, and because the principle of constant proportions states that the proportion of chlorinity to salinity is constant (that is, the ratio of various salts in seawater is the same in samples from many places regardless of how salty the water is), marine chemists have devised the following formula to determine salinity: *Salinity in $^o/_{oo}$ = 1.80655 x chlorinity in $^o/_{oo}$.* Chlorinity is about 19.4 $^o/_{oo}$, so salinity is around 35 $^o/_{oo}$.

Conductivity varies with the concentration and mobility of ions present, and with water temperature. Circuits in a conducting salinometer adjust for water temperature, convert conductivity to salinity, and then display salinity. Salinometers are calibrated against a sample of known conductivity and salinity. The best salinometers can determine salinity to an accuracy of 0.001%.

7. What factors affect seawater's pH? How does the pH of seawater change with depth? Why?

Seawater is slightly alkaline; its average pH is between 7.8 and 8.0. This seems odd because of the large amount of CO_2 dissolved in the ocean. If dissolved CO_2 combines with water to form carbonic acid (as noted above), why is the ocean mildly alkaline and not slightly acidic? When dissolved in water, CO_2 is actually present in several different forms. Carbonic acid (H_2CO_3) is only one of these. In water solutions some carbonic acid breaks down to produce the hydrogen ion (H^+), the bicarbonate ion (HCO_3^-), and the carbonate ion (CO_3^{2-}). This behavior acts to buffer the water, preventing broad swings of pH when acids or bases are introduced.

Though seawater remains slightly alkaline, it is subject to some variation. In areas of rapid plant growth, for example, pH will rise because CO_2 is used by the plants for photosynthesis. Because temperatures are generally warmer at the surface, less CO_2 can dissolve in the first place. Thus, surface pH in warm productive water is usually around 8.5.

At middle depths, and in deep water, more CO_2 may be present. Its source is the respiration of animals and bacteria. With cold temperatures, high pressure, and no photosynthetic plants to remove it, this CO_2 will lower the pH of water, making it more acid with depth. Thus, deep, cold seawater below 4,500 meters (15,000 feet) has a pH of around 7.5. A drop to pH 7 can occur at the deep ocean floor when bottom bacteria consume oxygen and produce hydrogen sulfide.

Thinking Critically

1. Where did the water of the ocean come from?

The volcanic venting of volatile substances from within the Earth -- outgassing -- gave rise to the present ocean. Much of this outgassed material was water vapor. As hot water vapor rose, it condensed into clouds in the cool upper atmosphere. Recent research suggests that millions of tiny icy comets colliding with the Earth may also have contributed to the accumulating mass of water vapor, this ocean-to-be.

The Earth's surface was so hot that no water could settle there, and no sunlight could penetrate the thick clouds. After millions of years the upper clouds cooled enough for some of the outgassed water to form droplets. Hot rains fell toward the Earth only to boil back into the clouds again. As the surface became cooler, water collected in basins and dissolved minerals from the rocks. Some of the water evaporated, cooled, and fell again. The world ocean gradually accumulated.

2. How are modern methods of determining salinity dependent on the principle of constant proportions?

Because of seawater's constancy of composition (the principle of constant proportions; Forchhammer's principle), we need to measure the concentration of only one major constituent to find the salinity of a water sample. The chloride ion is abundant and comparatively easy to measure, and it always accounts for the same proportion of dissolved solids (55.04 $\%_{oo}$), so marine chemists devised the concept of chlorinity to simplify measurement of salinity. Chlorinity is a measure of the total mass of halogen ions (the halogens are a chemical family that includes fluorine, chlorine, bromine, and iodine) in seawater.

3. What was the earthly origin of the sodium and chloride ions in common table salt?

Some of the ocean's solutes are hybrids of the two processes of weathering and outgassing. Table salt, or sodium chloride, is an example. The sodium ions come from the weathering of crustal rocks, while the chlorine ions come from the mantle by way of volcanic vents and outgassing from mid-ocean rifts.

4. Technically, there are no "salts" in seawater. How can that be?

An ion is an atom (or small group of atoms) that has an unbalanced electrical charge because it has gained or lost one or more electrons. Unlike the electron *sharing* found in covalently bonded molecules, the sodium atoms in NaCl have *lost* electrons, and chlorine atoms have *gained* them. The resulting ions are linked by the mutual attraction of their opposite electrical charges. The ions of sodium and chloride in NaCl are said to be held together by ionic bonds, electrostatic attraction that exists between ions that have opposite charge. When NaCl dissolves in water (Figure 7.3), the polarity of water reduces the electrostatic attraction (ionic bonding) between the sodium ion (Na^+) and chloride ion (Cl^-). This causes the sodium ion to separate from the chloride ion. The ions move away from the salt crystal, permitting water to attack the next layer of NaCl.

Note that NaCl does not exist as "salt" in seawater; its components are separated when salt crystals dissolve in water, but they are joined when crystals re-form as water evaporates.

Some other salts form ions in which groups of atoms bond together and carry an electrical charge. The sulfate ion (SO_4^{2-}), for example, carries two extra electrons. The charge on the overall ion is therefore -2.

5. How are seawater's conservative constituents different from its nonconservative constituents? Give an example of each.

Those constituents of seawater that occur in constant proportion or change very slowly through time are <u>conservative constituents</u>. Conservative elements have long residence times. Not surprisingly, these are the most abundant dissolved constituents of water -- the ones which constitute the bulk of the ocean's dissolved material. The inert gases dissolved in the ocean (and the water of the ocean itself) are also conservative constituents.

<u>Nonconservative constituents</u> are those substances dissolved in seawater which are tied to biological or seasonal cycles or to very short geological cycles. They have short residence times. Biologically important

nonconservative constituents include dissolved oxygen produced by plants, carbon dioxide produced by animals, silica and calcium compounds needed for plant and animal shells, or the nitrates and phosphates needed for production of protein and other biochemicals. Aluminum, with a residence time of only 600 years, is rapidly removed by absorption on clay sediment particles, so it is a nonconservative element. Aluminum is very rare in seawater (10 parts-per-billion).

6. *Which dissolved gas is present in the ocean in much greater proportion than in the atmosphere? Why the disparity?*

At the present time there is about 60 times as much carbon dioxide (CO_2) dissolved in the ocean as in the atmosphere.

Because CO_2 combines chemically with water to form a weak acid (H_2CO_3, carbonic acid), water can hold perhaps 1,000 times more carbon dioxide than either nitrogen or oxygen at saturation. Carbon dioxide is quickly used by marine plants, so dissolved quantities of CO_2 are almost always much less than this theoretical maximum.

7. *There's lots of oxygen in water (H_2O). Why can't fish breathe that? Why do they have to breathe oxygen dissolved in the water?*

The oxygen in the water molecule is tightly bound to the two hydrogen atoms by strong covalent bonds. Nothing happens during the passage of water across the gills of a fish to provide the large amount of energy needed to break those bonds and liberate the oxygen. Plants and plant-like photosynthesizers, however, can store enough sunlight to break the bonds to liberate the oxygen. (The hydrogen is used in electron transport reactions to provide energy build glucose in the photosynthetic process.) This oxygen can dissolve (usually as diatomic oxygen: O_2) and pass readily across gill membranes.

Thinking Analytically

1. *Imagine you have two small containers (say 10 milliliters each). Fill one with seawater and the other with pure water. Now, drop by drop, add dilute hydrochloric acid to each container and swirl the solution. After each drop, check the pH. What do you suppose a graph of pH vs. the number of drops would look like for pure water? For seawater? Explain the difference.*

Seawater acts as an effective buffer. A drop of dilute hydrochloric acid would be neutralized by reaction of the acid with the ions in solution. Another few drops would be affected the same way. Eventually the acid would overpower the buffering ability of the small volume of seawater, and the pH would drop rapidly. The curve would show no change with the first few drops, and then fall.

In contrast, the first drop of dilute hydrochloric acid place in pure water would reduce the pH by a small amount. Each succeeding drop would cause a proportional and linear decline in pH.

2. Why is the amount of dissolved chloride ion (Cl^-) a constant in the ocean, but the amount of dissolved carbon dioxide (CO_2) greatly variable?

The chloride ion is a conservative constituent of seawater – its residence time is very long and it is not tied to seasonal or biological cycles. Carbon dioxide, on the other hand, is in great demand by photosynthesizers as a source of carbon for the assembly of carbohydrate molecules. Carbon dioxide is a nonconservative element.

3. What is the salinity of a sample in parts-per-thousand if its chlorinity measures 13.4‰?

Review the equation on page 167. Plug in the numbers, and you'll find a salinity of $24.21‰$.

CHAPTER 8

CIRCULATION OF THE ATMOSPHERE

Reviewing What You've Learned

1. What is the composition of air?

The lower atmosphere is a nearly homogenous mixture of gases, most plentifully nitrogen (78.1%) and oxygen (20.9%). Other elements and compounds, as listed in Figure 8.1, make up less than 1% of its composition.

Air is never completely dry; water vapor, the gaseous form of water, can occupy as much as 4% of its volume. Sometimes liquid droplets of water are visible as clouds or fog, but more often the water is simply there, invisible, having entered the atmosphere from the ground, plants, and the sea surface.

2. Can more water vapor be held in warm air or cool air?

Warm air can hold more water vapor than cold air. Water vapor in rising, expanding, cooling air will often condense into clouds (aggregates of tiny droplets) because the cooler air can no longer hold as much water vapor. If rising and cooling continues, the droplets may coalesce into raindrops or snowflakes. The atmosphere will then lose water as precipitation, liquid or solid water that falls from the air to the Earth's surface.

3. Which is denser at the same temperature and pressure: humid air or dry air?

The temperature and water content of air greatly influence its density. Because the molecular movement associated with heat causes a mass of warm air to occupy more space than an equal mass of cold air, warm air is less dense than cold air. And contrary to what we might guess, humid air is *less* dense than dry air at the same temperature because molecules of water vapor weigh less than the nitrogen and oxygen molecules that the water vapor displaces.

4. What happens when air containing water vapor rises?

Air expands as it rises, and air cools as it expands. When air rises and cools, it cannot hold as much water in solution. The water comes out of solution and into suspension as visible droplets (clouds and fog). If the process continues, and atmospheric conditions are suitable, the droplets will coalesce (clump).

5. What causes precipitation (rain and snow)?

Water vapor in rising, expanding, cooling air will often condense into clouds (aggregates of tiny droplets) because the cooler air can no longer hold as much water vapor. If rising and cooling continues, the droplets may coalesce into raindrops or snowflakes. The atmosphere will then lose water as precipitation, liquid or solid water that falls from the air to the Earth's surface.

6. How many atmospheric circulation cells exist in each hemisphere?

There are three cells in each hemisphere: A Hadley cell between the equator and 30°, a Ferrel cell between 30° and 60°, and a polar cell above 60°.

7. Where does air rise, and fall, in each hemisphere. How does this movement affect weather?

Air rises at the equator and at the polar front (around 60°) in each hemisphere. Air falls at the junction between the Hadley and Ferrel cells (around 30°), areas characterized by deserts on land and high evaporation at sea. Air also falls near both poles.

8. What are the two kinds of large storms? How do they differ? How are they similar?

Extratropical cyclones form at a front between *two* air masses. Tropical cyclones form from disturbances within *one* warm and humid air mass. Each is characterized by strong winds, often accompanied by precipitation; each is a cyclone, a huge rotating system of low-pressure air in which winds converge and ascend.

9. What causes an extratropical cyclone? What happens in one?

Extratropical cyclones form at the boundary between each hemisphere's polar cell and its Ferrel cell -- the polar front. Because of the difference in wind direction in the air masses north and south of the polar front, the wave shape will enlarge, and a twist will form along the front. The different densities of the air masses prevent easy mixing, so the cold dense air mass will slide beneath the warmer lighter one. Formation of this twist in the northern hemisphere, as seen from above, is shown in Figure 8.19. The twisting mass of air becomes an extratropical cyclone.

Precipitation can begin as the circular flow develops. Precipitation is caused by the lifting, and consequent expansion and cooling, of the mass of mid-latitude air involved in the twist. As it rises and cools, this air can no longer hold all of its water vapor, so clouds and rain result. When *cold* air advances and does the lifting a "cold front" occurs. A "warm front" happens when *warm* air is blown on top of the retreating edge of cold air. The wind and precipitation associated with these fronts are sometimes referred to as frontal storms.

10. What triggers a tropical cyclone? From what is its great power derived? What causes the greatest loss of life and property when a tropical cyclone reaches land?

Unlike extratropical cyclones, these greatest of storms form within *one* warm, humid air mass between 10° and 25° latitude in both hemispheres. The origins of tropical cyclones are not well understood. A tropical cyclone usually develops from a small tropical depression. Tropical depressions form in easterly waves, areas of lower pressure within the easterly trade winds. Easterly waves, sometimes accompanied by a westward-moving line of thunderstorms, are thought to originate over a large, warm land mass -- indeed, the precursor storms that give rise to most of the destructive hurricanes that strike the East and Gulf coasts of the United States originate over western Africa.

A tropical cyclone is an ideal machine for "cashing in" water vapor's latent heat of evaporation, the source of the storm's vast energy. Warm, humid air forms in great quantity only over a warm ocean. As hot, humid tropical air rises and expands, it cools and is unable to contain the moisture it held when warm. Rainfall begins. Tremendous energy is released as this moisture changes from water vapor to liquid. Thus, solar energy ultimately powers the storm in a cycle of heat absorption, evaporation, condensation, and conversion of heat energy to kinetic energy.

Three aspects of a tropical cyclone can cause property damage and loss of life: wind, rain, and storm surge. But the most danger lies in storm surge, a mass of water driven by the storm. The low atmospheric pressure at the storm's center produces a dome of seawater that can reach a height of 1 meter (3.3 feet) in the open sea. The water height increases when waves and strong hurricane winds ramp the water mass ashore. If a high tide coincides with the arrival of all this water at a coast, or if the coastline converges (as is the case in Bangladesh at the mouths of the Ganges), rapid and catastrophic flooding will occur. Storm surges of up to 12 meters (40 feet) were reported at Bangladesh in 1970; 300,000 people drowned in less than 1 hour.

Thinking Critically

1. What factors contribute to the uneven heating of Earth by the sun?

Latitude is important. Sunlight striking polar latitudes spreads over a greater area, approaches the surface at a low angle favoring reflection, and filters through more atmosphere. Polar regions receive no sunlight at all during the depths of local winter. Contrast this to the tropical latitudes. The high solar angle in the tropics distributes the same amount of sunlight over a much smaller area, the more nearly vertical angle at which the light approaches means that it passes through less atmosphere and minimizes reflection. As you would expect, the tropics are warmer than the polar regions.

The seasons also play a role. The mounting angle you may have noticed in library reference globes indicates a tilt, the orbital inclination. The inclination of the Earth's axis causes the change of seasons and the Earth progresses through its year, facing each hemisphere alternately toward and away from the sun. At mid-latitudes the northern hemisphere receives about three times as much solar energy per day in June as it does in December.

The time of day is also important. The rotating Earth spins around the polar axis. This spin causes the daily rising and setting of the sun; morning and afternoon sunlight is not as strong as mid-day.

2. How does the atmosphere respond to uneven solar heating? How does the rotation of the Earth affect the resultant circulation?

Surface temperatures are higher at the equator than at the poles, and air can gain heat from warm surroundings. Since air is free to move over the Earth's surface, it would be reasonable to assume an air circulation pattern like the one that would develop in a room with a hot radiator at one end and a cold window at the other. In this ideal model, air heated in the tropics would expand and become less dense, rise to high altitude, turn poleward, and "pile up" as it converged near the poles. The air would then cool and contract by radiating heat into space, sink to the surface, and turn equatorward, flowing along the surface to the tropics to complete the circuit.

But this is not what happens. Global circulation of air is governed by two factors: uneven solar heating and the rotation of the Earth. The eastward rotation of the Earth on its axis deflects the moving air or water (or any moving object having mass) away from its initial course. To an earthbound observer, any object moving freely across the globe appears to curve slightly from its initial path. In the northern hemisphere, this curve is to the right (or clockwise) from the expected path; in the southern hemisphere, to the left (or counter-clockwise). This deflection is called the Coriolis effect in honor of

Gustave-Gaspard Coriolis, the French scientist who worked out its mathematics in 1835.

Coriolis effect and the movement of air from equator to pole (and back again) do not result in a single continuous loop flowing in each hemisphere. Because air loses heat the space as it moves at high altitude from near the equator toward the poles, some air falls toward the ground, "short circuiting" the flow. And because air gains heat from the Earth as it moves at low altitude from near the pole toward the equator, some air rises, again "short circuiting" the flow. The resulting 3-cell flow in each hemisphere is shown in Figure 8.13.

3. Why doesn't the ocean boil away at the equator, and freeze solid near the poles?

Because the atmosphere and the ocean move heat from the equatorial areas (where solar heat input exceeds heat outflow) to the polar regions (where heat outflow exceeds solar heat input). The physical movement of seawater (ocean currents) is responsible for about a third of the heat transport; the latent heat of evaporation of water (and carrying of water vapor by winds toward the polar regions) is responsible for about two-thirds. Review Figure 8.5a.

4. Describe the atmospheric circulation cells in the Northern Hemisphere. At what latitudes does air move vertically? Horizontally? What are the trade winds? The westerlies? Where are deserts located? Why? What is ocean surface salinity like in these desert bands?

The flow pattern mentioned above consists of large circuits of air is called an atmospheric circulation cells. A pair of these tropical cells exists, one on each side of the equator. They are known as Hadley cells. Look for them, along with the trade winds at their hearts, in Figure 8.13. A more complex pair of circulation cells operates at mid-latitudes in each hemisphere. Some of the air descending at 30° latitude turns poleward rather than equatorward. Before this air descends to the surface it is joined by air at high altitude returning from the north. As can also be seen in Figure 8.13, a loop of air forms between 30° and about 50°-60° of latitude. As before, the air is driven by uneven heating and influenced by the Coriolis effect. Surface air in this circuit is again deflected to the right, this time flowing from the west to complete the circuit. (This air is represented by the arrows labeled *westerlies*.) The mid-latitude circulation cells of each hemisphere are named Ferrel cells. They, too, can be seen in Figure 8.13.

Meanwhile, air that has grown cold over the poles begins moving toward the equator at the surface, turning to the west as it does so. At between 50° and 60° latitude in each hemisphere, this air has taken up enough heat and moisture to ascend. However, this polar air is more dense than the air in the adjacent Ferrel cell and does not mix easily with it. The unstable zone between these two cells generates most mid-latitude weather. At high altitude the ascending air from 50° - 60° latitudes turns poleward to complete a third circuit. These are the polar cells.

The desert bands are centered over those areas where air is descending. That air is dry (having lost moisture as it ascended elsewhere), and warms due to compressional heating as it nears the surface. Notice the position of the Sahara desert (northern Africa) relative to the 30° north latitude line. Following that line around the Earth would take you to the deserts of the American southwest, and the Gobi desert of Asia. The 30° south latitude line transects the Namib desert in southern Africa. As one might expect, ocean surface salinities in these areas are relatively high because of relatively warm temperatures (and consequent high evaporation) and low precipitation.

5. How are the geographic equator, meteorological equator, and ITCZ related? What happens at the ITCZ?

Sailors have a special term for the calm equatorial areas where the surface winds of the two Hadley cells converge: the doldrums. Scientists who study the atmosphere call this area the intertropical convergence zone, or ITCZ, to reflect the influence of wind convergence on conditions near the equator. Strong heating in the ITCZ causes surface air to expand and rise. The humid, rising, expanding air loses moisture as rain, some of which contributes to the success of tropical rain forests.

A consequence of the markedly different proportions of land to ocean surface in the two hemispheres is the position of the ITCZ. The convergence zone does not coincide with the geographical equator (0° latitude). Instead it lies at the meteorological equator (or *thermal equator*), an irregular imaginary line of thermal equilibrium between hemispheres situated about 5° north of the geographic equator. The positions of the meteorological equator and the ITCZ generally coincide, changing with the seasons, moving slightly farther north in the northern summer and returning toward the equator in the northern winter (Figure 8.13). Atmospheric and oceanic circulation in the two hemispheres is approximately symmetrical about the meteorological equator, not the geographical equator. Thus the doldrums, trade winds, horse latitudes, and westerlies shift north in the northern summer, and south in the northern winter.

6. What is a monsoon? How is monsoon circulation affected by the position of the ITCZ?

63

The intensity and location of <u>monsoon</u> activity depends on the position of the ITCZ. Note in Figure 8.16 that the monsoons follow the ITCZ south in the northern hemisphere's winter, and north in its summer.

7. If the Coriolis effect causes the clockwise deflection of moving objects in the Northern Hemisphere, why does air rotate counterclockwise around zones of low pressure in that hemisphere?

This apparent anomaly is caused by the Coriolis deflection of winds approaching the center of low pressure area from great distances. There is a *rightward* deflection of the *approaching* air. The edge spin given by this approaching air causes the storm to spin counterclockwise in the northern hemisphere.

Thinking Analytically

1. There is no such thing as "suck." Imagine the palm of your hand covering an open and empty peanut butter jar. Now imagine the air being pumped out of the jar. Your hand is not being "sucked" into the jar. Your hand is forced tightly onto the jar by differential pressure – the air pressure outside the jar is greater than the air pressure inside. Knowing the mass of a 1-square-centimeter column of air, calculate the total inward force exerted on the palm of your hand if it were covering a jar with a diameter of 7.5 centimeters and all the air could be removed from inside the jar.

On page 178 you learned that a column of air 1 centimeter square weighs 1.04 kilograms (2.3 pounds). The area of the empty jar's mouth is pi (3.14) x the square of its radius (14.06 square centimeters) = 44.2 square centimeters. Assuming a perfect vacuum could be obtained, multiplying this figure by the weight of one square centimeter yields a downward force of 46 kilograms, or 102 pounds. Air seems heavier than you may have thought!

2. Why does water drip from the bottom of your car when the air conditioner is being used on an especially humid day?

The cold part of the air conditioning apparatus – the part to which warm air loses its heat and becomes cool – is called the evaporator. Warm air can hold more water vapor than cool air (which is why fog sometimes forms when warm air is cooled by contact with a cold ocean). When humid air is cooled by rapid passage across the evaporator coils in the car's dashboard, water vapor condenses to form droplets on the metal coils and falls into a catch basin (or onto your feet if there's a leak). Water flows from the catch basin to a tube leading under the car.

3. Why is the longest day of the year almost never the hottest day of the year?

Because of thermal inertia. Water's extraordinarily high heat capacity allows the ocean to absorb an extraordinary amount of solar heat without rising much in temperature. It takes a long time for the ocean to warm up, so the hottest days in the Northern Hemisphere usually occur in August or early September, not in June. Likewise, the ocean gives up heat slowly, so the coldest days in this hemisphere are not in December, but in late January or early February. (The seasons are reversed in the Southern Hemisphere).

What do you think conditions on Earth would be like if the ocean were removed from its basins and sunlight could strike the deep sediments? How would temperatures and dates of hottest/coldest days differ from the present situation?

CHAPTER 9

CIRCULATION OF THE OCEAN

Reviewing What You've Learned

1. What forces are responsible for the movement of ocean water in currents?

Ocean currents are affected by two kinds of forces: the primary forces that start water moving and determine its velocity, and the secondary forces and factors that influence the direction and nature of its flow.

Primary forces -- those responsible for initial movement -- are thermal expansion and contraction of water, the stress of wind blowing over the water, and density differences between water layers.

2. What forces and factors influence the direction and nature of ocean currents?

Secondary forces and factors -- those influences that affect the direction and flow characteristics of a current -- are the Coriolis effect, gravity, friction, and the shape of the ocean basins themselves.

3. What is a gyre? How many large gyres exist in the world ocean? Where are they located?

Because of the Coriolis effect, northern hemisphere surface currents flow to the right of the wind direction. Southern hemisphere currents flow to the left. Intervening continents and basin topography often block continuous flow and help to deflect the moving water into a circular pattern. This flow around the periphery of an ocean basin is called a gyre.

There are six great current circuits in the world ocean, two in the northern hemisphere and four in the southern. They are shown in Figure 9.8. Five are geostrophic gyres: the North Atlantic Gyre, the South Atlantic Gyre, the North Pacific Gyre, the South Pacific Gyre, and the Indian Ocean Gyre. Though it is a closed circuit, the sixth and largest current is technically not a gyre, because it does not flow around the periphery of an ocean basin. The Antarctic Circumpolar Current, as this exception is called, flows endlessly eastward around Antarctica driven by powerful, nearly ceaseless westerly winds. This greatest of all surface ocean currents is never deflected by a continental mass.

4. What are water masses? Where are distinct water masses formed? What determines their relative position in the ocean?

A water mass is a body of water identifiable by its salinity and temperature (and therefore its density) or by its gas content or another indicator.

No matter at what depth they are located, the characteristics of each water mass have been determined by conditions of heating, cooling, evaporation, and dilution that occurred at the ocean *surface* when the mass was formed. The heaviest (and deepest) masses were formed by surface conditions that caused water to become very cold and salty. Water masses near the surface are warmer, less saline, and may have formed in warm areas where precipitation exceeded evaporation. Water masses at intermediate depths are intermediate in density.

Like air masses, water masses don't mix easily when they meet, but instead flow above or beneath each other, sorting themselves by density. Oceanographers name water masses according to their relative position. Water masses can be remarkably persistent and will retain their identity for great distances and long periods of time.

5. What drives the vertical movement of ocean water? What is the general pattern of thermohaline circulation?

The slow circulation of water at great depths is driven by density differences rather than by wind energy. Because density is largely a function of water temperature and salinity, the movement of water due to differences in density is called thermohaline circulation. Virtually the entire ocean is involved in slow thermohaline circulation, a process responsible for most of the vertical movement of ocean water.

Water sinks relatively rapidly in a small area where the ocean is very cold, rises much more gradually across a very large area in the warmer temperate and tropical zones, and slowly returns poleward near the surface to repeat the cycle.

More rapid and localized thermohaline currents exist. Some currents may move as rapidly as 60 centimeters per second (2 feet per second). These relatively fast currents are strongly influenced by bottom topography, and are sometimes called contour currents because their dense water flows around (rather than over) sea floor projections.

6. What methods are used to study ocean currents?

Traditional methods of measuring currents divide into two broad categories: the float method and the flow method. The float method depends on the movement of a drift bottle or other free-floating object. In the flow method the current is measured as it flows past a fixed object.

Surface currents can be traced with drift bottles or drift cards. These tools are especially useful in determining coastal circulation but provide no information on the path the drift bottle or card may have taken between its release and collection points. More elaborate drift devices can be tracked continuously by radio direction finders or radar. Deeper currents can also be surveyed by free-floating devices. The Swallow float (named after its developer) is used to detect the drift of intermediate water masses. Adjusted to descend to a specific density (and therefore depth), the Swallow float emits sonar "pings" as it drifts so that a tracking vessel can follow the movement of the water mass in which it is embedded.

Current meters, or flow meters, measure the speed and direction of a current from a fixed position. Most flow meters, such as the Ekman type shown in Figure 9.28d, use rotating vanes to measure current speed and a recording compass to measure direction. Bottom water movements are usually too slow to be measured by flow meters. Advances in electronics and computer design have made possible several new methods of study. One new device pioneered by the U. S. Office of Naval Research measures current speed by sensing the electromagnetic force generated by seawater as it moves in Earth's magnetic field. Buoys equipped with these sensors can record current speed and direction without dependence on delicate moving parts.

Sophisticated independent (autonomous) devices are now being deployed. The Sofar float seen in Figure 9.28c is an example. Shown being launched from a Woods Hole Oceanographic Institution research vessel, the probe will drop to a depth of 3,500 meters (11,500 feet) and produce a low-frequency tone once each day for tracking.

Thinking Critically

1. Why does water tend to flow around the periphery of an ocean basin? Why are western boundary currents the fastest ocean currents? How do they differ from eastern boundary currents?

The Coriolis effect influences any moving mass as long as it moves, so water in a gyre might be expected to curve to the center and stop. To understand why water continues to flow along the periphery of the gyre, imagine the forces acting on a surface water particle at 15° north latitude (as in Figure 9.6). Some flowing water will have turned to the right to form a hill of water near the center of the gyre -- it followed the rightward dotted-line arrow in the Figure. Why does the water now go straight west from point B without deflecting? Because, as Figure 9.7 shows, to turn further right the water would have to move uphill in defiance of gravity, but to turn left in response to gravity would defy the Coriolis effect. So the water continues westward, dynamically balanced between the force of gravity and Coriolis deflection.

Why should western boundary currents be concentrated and fast, and eastern boundary currents be diffuse and slow? One reason is the converging flow of the trade winds on either side of the equator. Water moved by the trades approaches the meteorological equator and is "shepherded" west where it piles up at the western edge of the basin before turning swiftly poleward. This concentration of water produces the poleward-moving western boundary currents. In contrast, the westerly winds of each hemisphere do not converge and water driven by them is not swept along a line of convergence. Coriolis deflection can therefore move some of the eastward-moving water equatorward before the basin's eastern boundary is reached.

A second reason is the rotation of the Earth itself. The hill described earlier is offset to the west because of the Earth's eastward rotation (in both hemispheres, of course), so water must squeeze closer to the ocean basin's western edge to pass around the hill at the western boundary. The combined effect on current flow is known as westward intensification.

2. What causes El Niño? How does an El Niño situation differ from normal current flow? What are the usual consequences of El Niño?

The trade winds normally drag huge quantities of water westward along the ocean's surface on each side of the equator, but sometimes the winds weaken and these equatorial currents crawl to a stop. Warm water that has accumulated at the western side of the Pacific—the warmest water in the world ocean—can then move to the east along the equator toward the coast of Central and South America. The eastward-moving warm water usually arrives near the South American coast around Christmas time, forming an El Niño event. The effects are felt not only in the Pacific; all ocean areas at trade wind latitudes in both hemispheres can be affected.

Normally, a current of cold water, rich in upwelled nutrients, flows north and west away from the South American continent (Figure 9.19a and b). When the propelling trade winds falter during an ENSO event, warm equatorial water that would normally flow westward in the equatorial Pacific backs up to flow east (Figure 9.19c and d). The normal northward flow of the cold Peru Current is interrupted or overridden by the warm water. Upwelling within the nutrient-laden Peru Current is responsible for the great biological productivity of the ocean off the coasts of Peru and Chile.

During major ENSO events, sea level rises in the eastern Pacific, sometimes by as much as 20 centimeters (8 inches) in the Galápagos. Water temperature also increases by up to 7°C (13°F). The warmer water causes more evaporation, and the area of low atmospheric pressure over the eastern Pacific intensifies. Humid air rising in this zone, centered some 2,000 kilometers (1,200 miles) west of Peru, causes high precipitation in normally dry areas. The increased evaporation intensifies coastal storms, and rainfall inland may be much higher than normal. Marine and terrestrial habitats and organisms can be affected by these changes.

3. What are countercurrents? Undercurrents? How is El Niño thought to be related to these currents?

Countercurrents flow on the surface near a main current, but in the opposite direction from the main current. Countercurrents can also exist beneath surface currents, and these are sometimes referred to as undercurrents.

A greatly strengthened equatorial countercurrent causes much of El Niño, but oceanographers have recently discovered that the Pacific Equatorial Undercurrent also increases greatly in volume during an ENSO event. The normal northward flow of the cold Peru Current is interrupted or overridden by the warm water. This current, rich in upwelled nutrients, is responsible for the great biological productivity of the ocean off the coasts of Peru and Chile. When the Peru Current slows, fish and seabirds dependent on the abundant life it contains die or migrate elsewhere. During major El Niño events, sea level and water temperature rise at the mid-Pacific's eastern boundary. The warmer water causes more evaporation and an area of low atmospheric pressure develops, centered some 2,000 kilometers (1,300 miles) west of Peru. Humid air rising in this zone causes high precipitation in normally dry areas. Marine and terrestrial habitats and organisms can be significantly impacted by these changes.

4. What is the role of ocean currents in the transport of heat? How can ocean currents affect climate? Contrast the climate of a mid-latitude coastal city at a western ocean boundary with a mid-latitude coastal city at an eastern ocean boundary.

Surface currents distribute tropical heat worldwide. Warm water flows to higher latitudes, transfers heat to the air and cools, moves back to low latitudes, absorbs heat again, and the cycle repeats. The greatest amount of heat transfer occurs at mid-latitudes where about 10^{15} (ten million billion) calories of heat are transferred each second -- more than a million times the power consumed by all the world's human population! The combination of water flow and heat transfer to and from water influences climate and weather.

Consider the influence of currents on two American cities at similar latitudes. Summer months in San Francisco (on the eastern boundary of an ocean basin) are cool, foggy, and mild; while Washington, D.C., on nearly the same line of latitude (but on a western boundary), is infamous for its August heat and humidity. Why the difference? Look at Figure 9.8 and follow the currents responsible. The California Current, carrying cold water from the north, comes close to the coast at San Francisco. Air normally flows clockwise in summer around an offshore zone of high atmospheric pressure. Wind approaching the California coast loses heat to the cold sea and comes ashore to chill San Francisco. Summer air often flows around a similar high off the East Coast (the Bermuda High). Winds approaching Washington, D.C., therefore blow from the south and east. Heat and moisture from the Gulf Stream contribute to the capital's oppressive summers. In winter Washington, D.C., is colder than San Francisco because westerly winds approaching Washington are chilled by the cold continent they cross, a land mass unable to retain much of its heat because the heat capacity of rock and dirt is less than that of water.

5. Can you think of ways ocean currents have (or might have) influenced history?

The voyage of the raft *Kon Tiki* indicates one possible answer. In 1947 Thor Heyerdahl, a Norwegian researcher and adventurer, decided to test the idea that the original Polynesians had drifted – intentionally or accidentally – from the mainland of South America to the Polynesian Islands, driven west with the prevailing winds and currents. To demonstrate this theory, he and five companions lashed together locally available balsa logs to form a raft and set out from a Peruvian seaport. After drifting aboard *Kon Tiki* (as the raft was named in honor of an Inca god), for about 8,000 kilometers (5,000 miles) and 101 days, Heyerdahl and his friends washed ashore on an atoll in the Tuamotu Islands east of Tahiti. Heyerdahl's theory gathered little support among scholars. He had proved that South Americans could have reached Polynesia, but did the Polynesians actually come from South America? As we saw in Chapter 2, modern anthropologists say they did not. Polynesian languages are similar to the languages of Southeast Asians, not South Americans. Studies of blood proteins have also shown parallels between Southeast Asian populations and the people of Polynesia. Perhaps most telling, Kon Tiki had to be towed 80 kilometers (50 miles) from shore; though the prevailing winds and currents flow from east to west, they do not commence until at least this distance from the mainland.

There are others ideas. Legends suggest the celebrated Celtic saint Brendan made an astonishing journey around the north Atlantic gyre, discovering the Azores and the coast of North America before returning to Ireland around 550 a.d. (His supposed safe completion of such a voyage in a small bent-branch-skeleton, hide-covered boat does not lend credence to the story.) If an earlier European discovery of North America had been accomplished, however, history would have been very different.

Pacific currents surely drove Polynesian navigators toward or away from distant destinations; the achievement despite currents of Easter Island and New Zealand by Polynesian navigators greatly influenced the history of those places.

Spectacular individual examples also come to mind. Read about the influence of currents on Captain William Bligh's open boat voyage to safety after the mutiny aboard HMS *Bounty*, or Sir Ernest Shackleton's astonishing open boat voyage in the howling West Wind Drift.

6. What holds up the thermocline? Wouldn't water slowly warm in an even gradient all the way to the bottom? (Hint: See Figure 9.24.)

The great quantities of dense water sinking at ocean basin edges must be offset by equal quantities of water rising elsewhere. Water sinks relatively rapidly in a small area where the ocean is very cold, but it rises much more gradually across a very large area in the warmer temperate and tropical zones. It then slowly returns poleward near the surface to repeat the cycle. The continual diffuse upwelling of deep water maintains the existence of the permanent thermocline found everywhere at low and mid-latitudes. This slow upward movement is estimated to be about 1 centimeter (½ inch) per day over most of the ocean. If this rise were to stop, downward movement of heat would cause the thermocline to descend and would reduce its steepness. In a sense, the thermocline is "held up" by the continual slow upward movement of water.

Thinking Analytically

1. Look again at Figure 9.7b, and consider the descending water. First, why does it stop descending? Second, which way is it most likely to go when it stops descending? (Hint: The Coriolis effect is greatest near the poles, and nonexistent at the equator.)

Density limits the water's descent – water beneath the pycnocline (thermocline) is more dense than the descending water, and not enough energy is available to allow mixing.

When the descending water begins to spread horizontally, its movement is (as always) subject to Coriolis effect. The gyral centers are in mid-latitudes, and the water would move outward toward the right (when viewed from above) as it spreads.

2. *Calculate the time it would take for an ideally situated rubber duck to make a loop of the North Atlantic. Now calculate the time it would take for an ideally situated rubber duck to make a loop of the North Pacific. Why the difference?*

For the Atlantic, begin with Figure 9.9 on page 206. If we assume "ideally situated" to be ideal for a high-speed circuit, we'll place the duck just off Cape Hatteras in a part of the Gulf Stream traveling at 2 meters per second. Now let's hope the duck doesn't get trapped in an eddy, but heads straight toward Ireland. The Stream slows as it moves into the mid-Atlantic, but average northeastern speed will be a bit less than 1 meter per second (about 2 miles per hour). Time to the British Isles will be about 100 days. The leisurely Canary Current would move the duck off western Africa in about another 90 days, and the circuit would be complete in about 10 months.

The Pacific is much more difficult to estimate, but preliminary estimates suggest a time at least twice as long. For one thing, the Pacific is much larger; for another, the flow is not as orderly (see Figure 9.8b). In the *Hansa Carrier* incident referred to in your text (pages 222-224), the lost shoes eddied around in the northern Pacific for a few years before entraining south and encountering beaches.

CHAPTER 10

WAVES

Reviewing What You've Learned

1. Draw an ocean wave and label its parts. Include a definition of wave period.

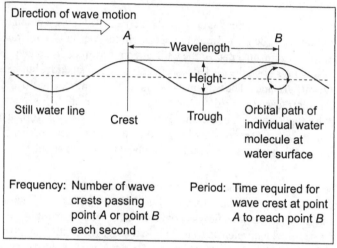

Frequency: Number of wave crests passing point *A* or point *B* each second

Period: Time required for wave crest at point *A* to reach point *B*

2. Make a list of ocean waves by disturbing force and wavelength. What is restoring force? How is a capillary wave different from a gravity wave?

Wavelength, Disturbing Forces, and Restoring Forces for Ocean Waves			
Wave Type	Typical Wavelength	Disturbing Force	Restoring Force
Capillary wave	Up to 1.73 cm (0.68 in.)	Usually wind	Cohesion of water molecules
Wind wave	60-150 m (200 - 500 ft)	Wind over ocean	Gravity
Seiche	Large, variable, a function of ocean basin size	Change in atmospheric pressure, storm surge, tsunami	Gravity
Seismic sea wave (tsunami)	200 km (125 mi)	Faulting of seafloor, volcanic eruption, landslide	Gravity
Tide	Half earth's circumference	Gravitational attraction, rotation of Earth	Gravity

74

Restoring force is the dominant force trying to return the water surface to flatness after a wave has formed in it. If the restoring force of a wave were quickly and fully successful, a disturbed sea surface would immediately become smooth and the energy of the embryo wave would be dissipated as heat.

The restoring force for very small water waves -- those with wavelengths less than 1.73 centimeters (0.68 inches) -- is cohesion, the property that allows individual water molecules to stick to each other by means of hydrogen bonds. These capillary waves are transmitted across a puddle because cohesion, the force that makes the tea creep up on the sides of a teacup, tugs the tiny wave troughs and crests toward flatness. Capillary waves are the first waves to form when the wind blows. These small ripples are important in transferring energy from air to water to drive ocean currents (as we saw in Chapter 9) but are of little consequence in the overall picture of ocean waves because they are tiny and carry very little energy.

All waves with wavelengths greater than 1.73 centimeters depend on gravity to provide the restoring force. Like the spring weight moving up and down, gravity pulls the crests downward, but the momentum of the water causes the crests to overshoot and become troughs. The repetitive nature of this movement gives rise to the circular orbits of individual water molecules in an ocean wave. These larger waves are called gravity waves. Since the circular motion of water molecules in a wave is nearly friction free, gravity waves can travel across thousands of miles of ocean surface without disappearing, eventually to break on a distant shore.

3. What factors influence the growth of a wind wave? What is a fully-developed sea? Where would we regularly expect to find the largest waves? Are waves in fully-developed seas always huge?

Three factors affect the growth of wind waves. Wind must be moving faster than the wave crests for energy transfer from air to sea to continue, so the mean speed of the wind, or wind strength, is clearly important to wind wave development. A second factor is the length of time the wind blows, or wind duration. The third is the uninterrupted distance over which the wind blows without significant change in direction, the fetch.

A fully-developed sea is the maximum wave size theoretically possible for a wind of a specific strength, duration, and fetch. Longer exposure to wind at that speed will not increase the size of the waves.

The greatest potential for large waves occurs beneath the strong and nearly continuous winds of the West Wind Drift surrounding Antarctica. The early nineteenth-century French explorer of the South Seas Jules Dumont d'Urville encountered a wave train with heights estimated "in excess" of 30 meters (100 feet) in Antarctic waters, and in 1916 Ernest

Shackleton contended with occasional waves of similar size in the West Wind Drift during a heroic voyage to remote South Georgia Island in an open boat. Satellite observations have shown that wave heights to 11 meters (36 feet) are fairly common in the Antarctic Circumpolar Current!

4. What happens when a wind wave breaks? What factors affect the break? How are plunging waves different from spilling waves?

The process begins with the transition of a deep-water wave to an intermediate-depth water wave in water less than half a wavelength deep. When the water is less than 1/2 the wavelength in depth, the wave "feels" bottom. The circular motion of water molecules in the wave is interrupted. Circles near the bottom flatten to ellipses. The wave's energy must now be packed into less water depth, so the wave crests become peaked rather than rounded. Friction with the bottom slows the wave, but waves behind it continue toward shore at the original rate. Wavelength therefore decreases but period remains unchanged. The wave becomes too high for its wavelength, approaching the critical 1:7 ratio. As the water becomes even shallower, the part of the wave below average sea level slows because of friction with the bottom. When the wave was in deep water, molecules at the top of the crest were supported by the molecules ahead (thus transferring energy forward). But this is now impossible because the *water* is moving faster than the *wave*. As the crest moves ahead of its supporting base, the wave breaks. The break occurs at about a 3:4 ratio of wave height to water depth. (That is, a 3 meter wave will break in 4 meters of water.) The turbulent mass of agitated water rushing shoreward during and after the break is known as surf. The surf zone is the region between the breaking waves and the shore.

The break can be violent and toppling, leaving an air-filled channel (or "tube") between the falling crest and the foot of the wave. These plunging waves form when waves approach a steeply sloping bottom. A more gradually sloping bottom generates a milder spilling wave as the crest slides down the face of the wave.

5. Describe wave reflection, refraction, and diffraction.

A vertical barrier such as a seawall, large ship hull, or smooth jetty will reflect progressive waves with little loss of energy. If the waves approach the obstruction straight on, the reflected waves will move away from the obstruction in the direction from which they came. This wave reflection will cause interference in the form of vertical oscillations called standing waves.

What happens when a wave line approaches the shore at an angle, as it almost always does (for example, in Figure 10.20)? The line does not break simultaneously because different parts of it are in different depths of water.

The part of the wave line in shallow water slows down, but the attached segment still in deeper water continues at original speed, so the wave line bends, or refracts. The bend can be as much as 90° from the original direction of the wave train. This slowing and bending of waves in shallow water is called <u>wave refraction</u>. The refracted waves break in a line almost parallel to the shore.

<u>Wave diffraction</u> is the propagation of a wave around an obstacle. Unlike wave refraction (which depends on a wave's response to a change in velocity), wave diffraction depends on the interruption of the obstacle to provide a new point of departure for the wave. This is illustrated by the gap in the breakwater in Figure 10.22. Wave crests excite the water in the gap. Water moving in the gap generates smaller waves in the harbor, which radiate from the gap and disturb the quiet water. The diffracted waves are much smaller than those in the open ocean, but boats in the harbor might still be jostled.

6. What are internal waves? Are internal waves larger or smaller than most wind waves? Slower or faster?

Progressive waves can form at the boundary between water layers of different densities. These subsurface waves are called <u>internal waves</u>. As is the case with ocean waves at the air-ocean interface, internal waves possess troughs, crests, wavelength, and period.

Normal ocean waves move rapidly because the difference in density between air and ocean is relatively great. Internal waves usually move very slowly because the density difference between the joined media is very small. Internal waves occur in the ocean at the base of the pycnocline, especially at the bottom edge of a steep thermocline (as you saw in Figure 10.25a). The wave height of internal waves may be greater than 30 meters (100 feet), causing the thermocline to undulate slowly through a considerable depth. Their wavelength often exceeds 0.8 kilometers (0.5 mile); periods are typically 5 to 8 minutes. Oceanographers are not certain how these large slow-motion waves are generated, but wind energy, tidal energy, or ocean currents may be responsible.

7. What causes tsunami? Are all seismic sea waves tsunami? Are all tsunami seismic sea waves? How fast do tsunami travel? Do they move in the same way or at the same speed in a confining bay as they would in the open ocean?

Tsunami are displacement phenomena -- waves that occur when seawater is shoved aside. Tsunami caused by the sudden vertical movement of the Earth along faults (the same forces that cause earthquakes) are properly called seismic sea waves. Landslides, icebergs falling from glaciers, volcanic eruptions, and other direct displacements of the water surface, can also cause tsunami. Note that all seismic sea waves are tsunami, but not all tsunami are seismic sea waves. The velocity of a tsunami is given by the formula for the speed of a shallow water wave: $C = \sqrt{gd}$. Solving for (C, celerity) shows that the wave would move at 212 meters per second (472 mph). At this speed a seismic sea wave will take only about five hours to travel from Alaska's seismically active Aleutians to the Hawaiian Islands.

A ship on the open ocean that encounters a tsunami with a 16 minute period would rise slowly and imperceptibly for about 8 minutes to a crest only 0.3 to 0.6 meters (1 or 2 feet) above average sea level, and then ease into the following trough 8 minutes later. With all the wind waves around, such a movement would not be noticed. But when a tsunami reaches shore, trouble often ensues.

8. Are tsunami ever dangerous if encountered in the open sea? What happens when they reach shore?

Seismic sea waves, as noted above, are rarely a problem in the open sea. As the seismic sea wave crest approaches shore, however, the situation changes rapidly and often dramatically for the worse. The period of the wave remains constant, velocity drops, and wave height greatly increases. As the crest arrives at the coast, observers would see water surge ashore in the manner of a very high, very fast tide. In confined coastal waters relatively close to their point of origin, tsunami can reach a height of 30 meters (100 feet). The wave is a fast on-rushing flood of water, not the huge plunging breaker of popular folklore. The dramatic photographs in Figures 10.32 through 10.35 give some indication of the difficulties involved!

Thinking Critically

1. Though they move across the deepest ocean basins, seiches and tsunami are called shallow-water waves. How can this be?

The wavelength of seismic sea waves usually exceeds 100 kilometers (62 miles). No ocean is 50 kilometers (31 miles) deep (half the wavelength), so seiches, seismic sea waves, and tides are always in water that to them is shallow or intermediate in depth, their huge orbit circles flattening against a distant bottom always less than half a wavelength away.

2. How do particles move in an ocean wave? How is that movement similar or different from the movement of particles in a wave in a spring or a rope? How does this relate to a stadium wave -- a waveform made by sports fans in a circular arena?

The nearly friction-free transfer of energy from water particle to water particle in these circular paths, or orbits, transmits wave energy across the ocean surface and causes the wave to move. This kind of wave is known as an orbital wave -- a wave in which particles of the medium (water) move in closed circles as the wave passes. Orbital ocean waves occur at the boundary between two media (between air and water), or between layers of water of different densities. Particles in a rope or a spring move only side to side (or forward and back), not in circular orbits. The wave in a rope or spring is thus not an orbital wave. But because the wave in all these waves moves forward, they are all known as progressive waves.

A stadium wave is a particularly effective demonstration of the fallacy of the wave illusion -- the tendency we have to think of a wave as a physical object. Participants in a stadium wave need only stand and sit at the appropriate time to propagate the wave. The fans don't move laterally -- they certainly don't leave their seats and run around the stadium to make the wave go. Yet the wave appears to be a thing, an illusion all progressive waves share.

3. What is the general relationship between wavelength and wave speed? How does water movement in a wave change with depth? How does water movement in a wave change with depth?

In general, the longer the wavelength of a wave, the faster the wave energy will move through the water. For deep-water waves this relationship is shown in the formula

$$C = \frac{L}{T}$$

where C represents celerity (speed), L is wavelength, and T is time, or period (in seconds). Figure 10.7 shows the relationship between wavelengths of deep-water waves and their velocity and period.

The action of shallow-water waves is described by a different equation, which may be written as

$$C = \sqrt{gd} \quad \text{or} \quad C = 3.1\sqrt{d}$$

where C is celerity, or speed, (in meters per second), g the acceleration due to gravity (an average of 9.8 meters per second per second), and d the depth of water in meters. The *period* of a wave remains unchanged regardless of the depth of water through which it is moving, but as deep-water waves enter the shallows and feel bottom their velocity is reduced and their crests "bunch up," so their wavelength shortens.

Deep water and shallow water waves can coexist at the same point offshore. For example, imagine two wave trains traveling over a shoal 50 feet deep. Waves with 20-foot wavelengths would be in deep water

$(20 \times 2 = 40 < 50)$. Waves with 60-foot wavelengths would be in shallow water $(60 \times 2 = 120 > 50)$.

4. How can a rogue wave be larger than the theoretical maximum height of waves in a fully-developed sea?

A rogue wave is a rare confluence of crests that converge from different directions at a single spot. The sudden interaction can lead to constructive interference, a situation in which wave energy is additive. The rogue is not a progressive wave, and thus is not limited to the theoretical maximum height in a fully-developed sea. Figure 10.15 shows a simple example of the operation of constructive interference.

5. How is a progressive wave different from a standing wave? Must standing waves only be orbital waves, or can standing waves also form in shaken ropes or pushed-and-pulled springs.

Standing waves take the form of vertical oscillations; progressive waves, as the name implies, appear to move forward. Standing waves can easily be induced in shaken ropes and pushed-and-pulled springs. To try making a standing wave with a rope, tie one end to a post 10 feet away, then jerk the end left and right rapidly. Experiment with the frequency. After a while, you should see a place develop at the center of the rope in which there is no side-to-side movement. Places in the rope 1/4 and 3/4 of the way to the post will be swinging in opposite directions with each cycle. That's a standing wave.

6. How can large waves generated by a distant storm arrive at a shore first, to be followed later by small waves.

It seems reasonable to assume that waves arriving at a distant shore from a big storm would increase in size, reach a peak, and then die away. But that's not what happens. Remember, wave speed is proportional to wavelength. The largest waves generated in a fetch will move quickly toward the shore; the smaller waves will move more slowly. So, at one time it can be "Lake California," but an hour later the surf can be roaring! The best time for surfing (or just wave watching) is during those first couple of hours before the smaller waves have arrived to begin complex resonance and interference patterns. Surfers always seem to have a network to telegraph these oceanographic observations to one another – the word goes out and the waves get crowded. The Internet has helped immeasurably, as you will discover when you go to the hyperlinks for this chapter in the book's website.

Thinking Analytically

1. *What is the celerity (speed) of a deep water wave having a period of 10 econds? What is its wavelength?*

Use the nomogram in Figure 10.7 on page 233 to make estimates: the speed is around 16.4 meters per second; wavelength is around 165 meters. Now solve the equations on pages 232-234 for C and L and plug in the numbers. Do your calculations confirm your estimates?

2. *Assuming the alarm was raised immediately, how much tsunami warning would residents of southeastern Japan have if a strong earthquake stuck the Peru-Chile trench?*

Look again at Figure 10.31 on page 249. The transit time in 1960 was 22.5 hours, plenty of time to mobilize residents to shut down equipment and move to higher ground. Solving the equation at the bottom of page 249 for C indicates the great speed at which tsunami travel. Still, the Pacific is the world's largest feature, and it takes nearly a day for a wave traveling at that speed to traverse it. Best not to waste any time, though.

3. *What is the wavelength of a typical storm surge?*

Technically a storm surge is not a wave because it is only a crest; so wavelength and period cannot be assigned to it.

4. *Can deep-water waves and shallow-water waves exist at the same point offshore (that is, in the same depth of water)?*

Yes. Look at Figure 10.20a on page 243. Find the line that says "Waves begin to 'feel bottom' here..." Waves of the wavelength drawn in this diagram become shallow water waves at that point. Out where "Direction of progress" is written, these waves would be deep-water waves. But waves with very long wavelengths could *already* be in water less deep than half their wavelength (that is, in shallow water) at the position of the "Direction of progress" line. So, halfway between those two lines at a single point over the seabed, one wave would be in deep water, and the other in shallow water. If this doesn't click in your mind, re-read the beginning of the section on deep water and shallow water waves on page 232 once more before giving up.

CHAPTER 11

TIDES

Reviewing What You've Learned

1. What causes the rise and fall of the tides? What celestial bodies are most important in determining tides? Are there such things as "tidal waves?"

Tides are caused by a combination of the gravitational force of the moon and sun and the motion of the Earth in its orbit. With a wavelength that can equal half Earth's circumference, tides are the longest of all waves. Unlike the other waves we have met, these huge shallow-water waves are never free of the forces that cause them, and so act in unusual but generally predictable ways.

When river mouths are exposed to very large tidal fluctuation, a tidal bore will sometimes form. Here is a form of tide-generated (tidal) wave—a steep wave moving upstream generated by the arrival of the tide crest in an enclosed river mouth. The funnel shape of the estuary confines the tide crest and forces it to move inland at a speed that exceeds the theoretical shallow water wave speed for that depth. The forced wave then breaks to form a spilling wave front that moves upriver.

Another candidate for "tidal wave" might be the crest of the tide wave itself. Don't confuse "tidal wave" with the earthquake- or displacement-derived terms "seismic sea wave" or "tsunami."

2. What is a high tide and a low tide? A spring tide and a neap tide? A tidal bore?

The moon's (and sun's) gravity pulls water toward points on the Earth beneath those bodies -- the tractive forces you read about in Figure 11.5. Inertia flings water toward points opposite those places. Water directly beneath the moon (and sun), and at spots on the Earth directly opposite those bodies, "bulge" with water. Figure 11.6 shows the result (greatly exaggerated in the figure) for the moon. Bulges have formed beneath the moon's position and at a point on Earth directly opposite that position. The bulges are the crests of the planet-sized waves that cause <u>high tides</u> as the Earth rotates. <u>Low tides</u> correspond to the troughs.

The large tides caused by the linear alignment of sun, Earth, and moon are called <u>spring tides</u>; these times of very high high tides and very low low tides occur at two-week intervals corresponding to new and full moon. (Please note that spring tides don't happen only in the spring of the year!) <u>Neap tides</u> alternate at two-week intervals; they occur when the Earth, moon, and sun form a right angle. During times of neap tides high tides are not very high, and low tides are not very low. Please see Figure 11.12 for a pictorial representation of neap tides and spring tides.

3. How does the equilibrium theory of tides differ from the dynamic theory?

Newton's gravitational model of tides—the <u>equilibrium theory</u>—deals primarily with the position and attraction of Earth, moon, and sun, and does not factor in the influence on tides of ocean depth or the positions of continental landmasses. The equilibrium theory would accurately describe tides on a planet uniformly covered by water. A modification proposed by Pierre-Simon Laplace about a century later—the <u>dynamic theory</u>—takes into account the speed of the long-wavelength tide wave in relatively shallow water, the presence of interfering continents, and the circular movement or rhythmic back-and-forth rocking of water in ocean basins.

4. How does an astronomical tide differ from a meteorological tide? Are the tides separate and independent of each other?

As the term suggests, an <u>astronomical tide</u> is caused by alignments of the sun and moon, and the dynamic effects of fluid motion. But sometimes a tide crest won't arrive at a predicted time or attain its predicted height along a coast.   A strong, steady wind onshore or offshore, or the low pressure beneath a tropical cyclone, can affect tidal height and the arrival time of the crest. Weather-related alterations are sometimes called <u>meteorological tides</u> after their origin.

Thinking Critically

1. Though they move through all the ocean, tides are referred to as shallow-water waves. How can that be?

Remember that tides are a form of wave. The crests of these waves— the tidal bulges—are separated by a distance of half of Earth's circumference (see again Figure 11.7). In the equilibrium model, the crests would remain stationary, pointing steadily toward (or away from) the moon (or sun) as Earth turned beneath them. They would appear to move across the idealized water-covered Earth at a speed of about 1,600 kilometers (1,000 miles) per hour. But how deep would the ocean have to be to allow these waves to move freely? For a tidal crest to move at 1,600 kilometers per hour, the ocean would have to be 22 kilometers (13.7 miles) deep. As you may recall, the average depth of the ocean is only 3.8 kilometers (2.4 miles). So, tidal crests (tidal bulges) move as <u>forced waves</u>, their velocity determined by ocean depth.

2. *What are the most important factors influencing the heights and times of tides? What tidal patterns are observed? Are there tides in the open ocean? If so, how do they behave?*

Tidal range (high-to-low-water height difference) varies with basin configuration and location. In small areas such as lakes, the tidal range is very small. In larger enclosed areas such as the Baltic or Mediterranean seas, tidal range is also moderate. Tidal range is not the same over a whole ocean basin -- it varies from the coast to the centers of oceans. Tides near the centers of ocean basins exist, but they tend to be small in magnitude. The largest tidal ranges occur at the edges of the largest ocean basins, especially in bays or inlets that concentrate tidal energy because of their shape. Tidal range at the apex of a funnel-shaped sea, gulf, or bay can often be extreme. In the Bay of Fundy near Moncton, New Brunswick (Canada), tidal range is especially wide: up to 15 meters (50 feet) from highs to lows! The northern reaches of the Sea of Cortez east of Baja California have a tidal range of about 9 meters (30 feet). Tide waves sweeping toward the narrow southern end of the North Sea can build to great heights along the southeast coast of England and the north coast of France.

The shape of the basin also has a strong influence on the patterns of tides. Because of basin resonances, some coastlines experience semidiurnal (twice daily) tides: two high tides and two low tides of nearly equal level each lunar day. Others have diurnal (daily) tides: one high and one low. The tidal pattern is called mixed if successive high tides or low tides are of significantly different heights through the cycle.

3. *How does the latitude of a coastal city affect the tides there -- or does it?*

It the Earth were uniformly covered by the ocean with no emergent land and a smooth seabed, the greatest tidal ranges would be confined to the latitudes between the greatest northerly and southerly movements of the moon and the sun (about 28° for the moon, 23° for the sun). Locations poleward of those lines would experience progressively less extreme tides. But basin resonances and the resulting amphidromic points complicate this simple model. As things stand, there is little relation between latitude and tidal height. For example, note that the tides at south Pacific node near Tahiti are very mild (See figure 11.16). Tides in the Aleutian Islands of western Alaska, high in the north, can be extreme. Basin shape, discussed in the previous question, also plays a role.

4. *From what you learned about tides in this chapter, where would you locate a plant that generated electricity from tidal power? What would be some advantages and disadvantages of using tides as an energy source?*

Tidal power has many advantages: Operating costs are low, the source of power is free, and no carbon dioxide or other pollutants are added to the atmosphere. But even if tidal power stations were built at every appropriate site worldwide, the power generated would amount to less than 1% of current world needs. And, of course, this method of power generation is not free of trade-offs. The dam and electrical generators can be damaged by storms, and the large finely made metal valves and vanes at the heart of the plant are easily corroded by seawater. Computer simulations have suggested that installing a dam would change the resonance modes of a bay or estuary and therefore the height of the tide wave. Studies also suggest that sensitive plankton and benthic marine life would be disrupted and even that increased tidal friction would cause a minute decrease in the rate of Earth's rotation.

Though residents and mariners would vigorously oppose such a project, a tidal dam beneath San Francisco's Golden Gate Bridge could generate more electrical power than any existing tidal generating station. It has the prime requisites: The narrow entrance to the Bay, combined with the large volume of water passing back and forth through the strait every day, would make such a site nearly ideal for this purpose. The esthetic, navigational, and structural difficulties presented by such a project mitigate very heavily against its successful completion, however.

Thinking Analytically

1. If you live near a coast, find a source of local tidal data (newspaper, Internet site, pamphlet from bait shop, etc.). Plot the rise and fall of the tides for two weeks. Note the cycle, and point out spring and neap tides. How does the tide correlate with the position of the moon and sun?

This one is up to you.

2. You know that tides always act as shallow water waves. What is the speed of a tide wave in the open ocean? Assume the average depth of the ocean is 4,000 meters.

The formula for speed (celerity) of a shallow water wave is

$$C = \sqrt{gd} \quad \text{or} \quad C = 3.1\sqrt{d}$$

$3.1 \times 63.25 = 196$ meters per second, or 438 miles per hour.

CHAPTER 12

COASTS

Reviewing What You've Learned

1. How is an erosional coast different from a depositional coast?

If a coast is essentially in almost the same condition as it was when sea level stabilized after the last ice age, it is called an <u>erosional coast</u>. Erosional coasts are young coasts in which terrestrial influences (that is, processes that occur at the boundary between land and air) dominate. If the coast has been significantly changed by wave action and other marine processes since sea level stabilized, it is termed a <u>depositional coast</u>. Depositional coasts are usually older than erosional coasts -- that is, they have been exposed to marine action for a longer time. Depositional coasts retain little (if any) evidence of the non-marine processes that produced them. Some coasts show both characteristics -- a single long shoreline can consist of both erosional and depositional coasts.

2. What features would you expect to see along an erosional coast? A depositional coast? How long would you expect the features to last?

<u>Erosional coasts</u> are often rough and irregular. The ocean has not had time to modify the terrestrial features provided by changes in sea level, the scouring of glaciers, deposition of sediment at the mouths of rivers, volcanic eruptions, or the movement of the Earth along faults. We could see sunken river valleys; deep, narrow embayments known as fjords; deltas; and coasts pocked by volcanic craters and lava flows.

<u>Depositional coasts</u> are those coasts that have been significantly changed by wave action and other marine processes after sea level stabilized. Land erosion and marine erosion both work to change a erosional coast to a depositional coast. Depositional coasts are shaped and attacked from the land by stream erosion, the abrasion of wind-driven grit, the alternate freezing and thawing of water in rock cracks, the probing of plant roots, glacial activity, rainfall, dissolution by acids from soil, and slumping. Erosive forces can produce a wave-cut shore showing some or most of the features illustrated in Figure 12.3. The most familiar feature of a depositional coast is the beach. A beach is a zone of unconsolidated (loose) particles that covers part or all of a shore.

As with all coastal situations, change is the dominant condition in both erosional and depositional coasts.

3. What are some of the features of a sandy beach? Are they temporary or permanent? Is there a relationship between wave energy on a coast and the size (or slope, or grain size) of beaches found there?

The features found on depositional coasts are usually composed of sediments rather than solid rock. Beaches are the dominant form. Most beach sediment reaches the coast in rivers. Not all the incoming sediment joins the longshore drift; some of the fine particles moved by rivers will stay in suspension long enough to be transported to the outer continental shelf and beyond. If the rate of deposition of larger particles exceeds the ability of the longshore transport system to remove and distribute the material along the coast, the sediments may build up at the river mouth to form a fan called a delta.

The landward limit of a beach may be vegetation, a sea cliff, relatively permanent sand dunes, or construction such as a seawall. The seaward limit occurs where sediment movement on- and offshore ceases -- a depth of about 10 meters (33 feet) at low tide. Such places include the calm spots between headlands, shores sheltered by offshore islands, and regions with usually quiet surf. Sometimes the sediment is transported a very short distance -- grit may simply fall from the cliff above and accumulate at the shoreline -- but more often the sediment on a beach has been moved for long distances to its present location.

Wherever they are found, deltas and beaches are in a constant state of change. They may be thought of as rivers of sand -- zones of continuous sediment transport.

High-energy beaches tend to look different than low-energy beaches; fine-grain beaches different from coarse-grain ones. On fine-grain beaches, the ability of small sharp-edged particles to interlock discourages water from percolating down into the beach itself, so water from waves runs quickly back down the beach carrying surface particles toward the ocean. This process results in a very gradual slope. Broad flat beaches also have a large area on which to dissipate wave energy and can provide a calm environment for the settling of fine sediment particles. In contrast, coarse particles (gravel, pebbles) do not fit together well and readily allow water to drain between them. Onrushing water disappears into a beach made of coarse particles, so little water is left to rush down the slope, thereby minimizing the transport of sediments back to the ocean. Thus larger particles tend to build up at the back of the beach, increasing its steepness.

4. How are deltas classified? Why are there deltas at the mouths of the Mississippi and Nile Rivers, but not at the mouth of the Columbia River?

The shape of a delta represents a balance between the accumulation of sediments and their removal by the ocean. For a delta to maintain its size or grow, the river that feeds the delta must carry enough sediment to keep marine processes in check. The combined effects of waves, tides, and river flow determine the shape of a delta. In 1975, William Galloway, a geologist at the University of Texas, classified deltas by the relative influence of those three factors (see Figure 12.24). Note that any delta strongly affected by only one of the three factors is placed at an apex of the triangle. A delta with a balance of forces is placed near the middle of the triangle. River-dominated deltas are fed by a strong flow of fresh water and continental sediments, and form in protected marginal seas. In tide-dominated deltas, freshwater discharge is overpowered by tidal currents that mold sediments into long islands parallel to the river flow and perpendicular to the trend of the coast. Wave-dominated deltas are generally smaller than either tide- or river-dominated deltas and have a smooth shoreline punctuated by beaches and sand dunes. Instead of a bird's-foot pattern of distributaries, a wave-dominated delta will have one erosional exit channel.

Deltas do not form at the mouth of every sediment-laden river. A broad continental shelf must be present to provide a platform on which sediment can accumulate, and, as befits a erosional coast, marine processes must not dominate -- that is, tidal range should be low, and waves and currents generally mild. Deltas are most common on the low-energy shores of enclosed seas where the tidal range is not extreme, and along the tectonically stable trailing edges of some continents. The largest deltas are those of the Gulf of Mexico (the Mississippi), the Mediterranean Sea (the Nile), the Ganges-Brahamaputra river system in the Bay of Bengal, and the huge deltas formed by the rivers of China that empty into the South China Sea. Deltas tend not to form at West Coast river mouths because the continental shelf is narrow, river flow is generally low (except for the Columbia River), and beaches are usually high in wave energy.

5. What is a coastal cell? Where does sand in a coastal cell come from? Where does it go?

The natural sector of a coastline in which sand input and sand outflow are balanced may be thought of as a coastal cell. The main features of such a cell are illustrated in Figure 12.17. The size of coastal cells varies greatly. They are often very large along the relatively smooth, tectonically passive trailing edges of continents; coastal cells along the southeast coast of the United States, for example, are hundreds of kilometers long. On the active leading edge of the continent they are smaller. Four cells exist in the 360 kilometers (225 miles) between southern California's Point Conception and the Mexican border. Each terminates in a submarine canyon at the downcoast end (see again Figure 12.17).

Sand enters the cell with river runoff, travels the length of the cell by longshore transport, and exits at the mouth of a submarine canyon.

88

6. How are estuaries classified? Upon what does the classification depend? Why are estuaries important?

Three factors determine the characteristics of estuaries: the shape of the estuary, the volume of river flow at the head of the estuary, and the range of tides at the estuary's mouth. The mingling of waters of different densities, the rise and fall of the tide, and the variations in river flow along with the actions of wind, ice, and Coriolis' effect guarantee that patterns of water circulation in an estuary will be complex.

Estuaries are classified by their circulation patterns. The simplest circulation patterns are found in salt wedge estuaries, which form where a rapidly flowing large river enters the ocean in an area where tidal range is low or moderate. The exiting fresh water holds back a wedge of intruding seawater. Note that density differences cause fresh water to flow over salt water. The seawater wedge moves seaward at times of low tide or strong river flow, and returns landward as the tide rises or when river flow diminishes. Some seawater from the wedge joins the seaward-flowing fresh water at the steeply sloped upper boundary of the wedge, and new seawater from the ocean replaces it. Nutrients and sediments from the ocean can enter the estuary in this way. Examples of salt wedge estuaries are the mouths of the Hudson and Mississippi rivers.

A different pattern occurs where the river flows more slowly and tidal range is moderate to high. As their name implies, well-mixed estuaries contain differing mixtures of fresh and salt water through most of their length. Tidal turbulence stirs the waters together as river runoff pushes the mixtures to sea. The mouth of the Columbia River is an example.

Deeper estuaries exposed to similar tidal conditions but greater river flow become partially-mixed estuaries. Partially-mixed estuaries share some of the properties of salt wedge and well-mixed estuaries. England's Thames River, San Francisco Bay, and Chesapeake Bay are examples.

Reverse estuaries can form along arid coasts when rivers cease to flow. The evaporation of seawater in the uppermost reaches of these estuaries will cause water to flow from the ocean into the estuary, producing a gradient of increasing salinity from the ocean to the estuary's upper reaches. Reverse estuaries -- sometimes called lagoons -- are common on the Pacific coast of Mexico's Baja peninsula and along the U. S. Gulf coast.

As we will see in Chapter 16, estuaries often support a tremendous number of living organisms. The easy availability of nutrients and sunlight, protection from wave shock, and presence of many habitats permit the growth of many species and individuals. Estuaries are frequently nurseries for marine animals; several species of perch, anchovy, and Pacific herring take advantage of the abundant food in estuaries during their first weeks of life. Unfortunately for their inhabitants, estuaries are in high demand for development into recreational resources and harbors. Estuaries have become the most polluted of all marine environments.

7. Compare and contrast the Pacific, Atlantic, and Gulf coasts of the United States.

The West Coast is an actively rising margin on which volcanoes, earthquakes, and other indications of recent tectonic activity are easily observed. West Coast beaches are typically interrupted by jagged rocky headlands, volcanic intrusions, or the effects of submarine canyons. Most of the sediments on the West Coast originated from erosion of relatively young granitic rocks of the coastal mountains. The particles of quartz and feldspar that comprise most of the sand were transported to the shore by flowing rivers. The volume of sedimentary material transported to west coast beaches from inland areas greatly exceeds the amount originating at the coast itself. Because West Coast beaches are usually high in wave energy, deltas tend not to form at West Coast river mouths. The predominant direction of longshore drift is to the south.

The East Coast is a passive margin, tectonically calm and subsiding because of its trailing central position on the North American Plate. Subsidence along the coast has been considerable over the last 150 million years, and a deep layer of sediment built up offshore, material that produced the ancestors of today's barrier islands. Relatively recent subsidence has been more important in shaping the present coast, however. Coastal sinking and rising sea level have combined to submerge some parts of the East Coast at a rate of about 1/3 meter (1 foot) per century. This process has formed the huge flooded valleys of Chesapeake and Delaware Bays, the landward-migrating barrier islands, and the shrinking lowlands of Florida and Georgia. Rocks to the north (in Maine, for example) are among the hardest and most resistant to erosion of any on the continent, so beaches are uncommon in Maine. But from New Jersey southward the rocks are more easily fragmented and weathered, and beaches are much more common. As on the West Coast, sediments are transported coastward by rivers from eroding inland mountains, but the transported material is trapped in sunken estuaries and therefore plays a less important role on beaches. Eastern beaches are typically formed of sediments from nearby erosional shores, or from the shoreward movement of offshore deposits laid down when the sea level was lower. The amount of sand in an area thus depends in part on the resistance or susceptibility of nearby shores to erosion. Sand moves generally south on these beaches just as it does on the West Coast, but the volume of moving sand in the East is less.

The Gulf Coast experiences a smaller tidal variance and – at least between hurricanes -- a smaller average wave size than either the West or East Coasts. Reduced longshore drift and an absence of interrupting submarine canyons allow the great volume of accumulated sediments from the Mississippi and other rivers to form large deltas, barrier islands, and a long raised "super berm" that prevents the ocean from inundating much of this sinking coast.

Thinking Critically

1. In what other ways can coasts be classified? What are the advantages of classification? Are there disadvantages?

Coasts are hard to classify because many factors combine to shape them. Each classification of coasts is important because it provides a standard for reference and comparisons. In the 1830s English geologists first classified coasts by the type of rock or debris found on them, but this method was superficial and did not explain how the materials came to that location. A later scheme, one of the first to include elements of this century's newly emerging understanding of geology, was proposed in 1919 by geologist Douglas Johnson. Johnson's classification was based on whether a coast was *rising* or *falling* -- emerging (rising) coasts show an advancing coastline with the shore moving seaward, subsiding (falling) coasts have retreating coastlines with the shore moving inland. The underlying causes of coastal rise or fall were not then understood. We now know that some Alaskan and most Scandinavian coasts are rising in isostatic response to removal of the weight of overlying ice since the last ice age. A few Alaskan harbors have emerged far enough since the mid-1800s to make them dangerous to navigation. The tectonic movement of Earth's lithospheric plates also causes coasts to emerge or subside.

As we saw in Chapter 3, no area of geology was left undisturbed by the revelations of plate tectonics. In the 1960s geologists began to classify coasts according to tectonic position. *Active* coasts, near the leading edge of moving continental masses, were found to be fundamentally different from the more *passive* shores near trailing edges. The ages and characteristics of coasts are better understood by taking plate movements into account, as are the coasts' composition and physical shapes. But the slow forces of plate movement are frequently obscured by the more rapid action of waves, by erosion of the land, and by the transport of sediments. Coasts are modified by small-scale (erosional/depositional) as well as large-scale (tectonic) processes.

2. What two processes contribute to longshore drift? What powers longshore drift? What is the predominant direction of drift on U. S. coasts? Why?

The net amount of sediment (usually sand) that moves along the coast, driven by wave action, is referred to as longshore drift. Longshore drift occurs in two ways: the wave-driven movement of sand along the exposed beach, and the current-driven movement of sand in the surf zone just offshore.

If sediments have accumulated to form a beach, water from breaking waves will usually rush up the beach at a slight angle (waves rarely approach the shore exactly at a right angle), but return to the ocean by running straight down hill under the influence of gravity. The millions of sand grains disturbed by the wave will follow the water's path, moving up the beach at an angle but retreating down the beach straight down the slope. Net transport of the grains is "longshore," parallel to the coast, *away* from the direction of the approaching waves.

Sediment is also transported in the surf zone in a longshore current. The waves breaking at a slight angle distribute a portion of their energy away from their direction of approach. This energy propels a narrow current in which sediment already suspended by wave action can be transported downcoast. The speed of the longshore current sometimes reaches almost 4 kilometers per hour (about 2.5 miles per hour).

Net sand flow along the U. S. East and West coasts is usually to the south because the waves that drive the transport system usually approach from the north, where storms most commonly occur.

3. What physical conditions might limit or encourage the development of a coral reef? An atoll?

Reef-building coral polyps secrete a cup-shaped calcium carbonate skeleton, which remains behind after the animal dies. The accumulation of skeletons gradually forms the reef. These corals grow best in brightly lighted seawater of normal (or slightly high) salinity about 5 to 10 meters (16 to 33 feet) deep. Fresh water is lethal to coral animals, even in relatively small dilutions. In ideal conditions they grow at a rate of about 1 centimeter (½ inch) per year. Coral reefs flourish in areas where offshore upwelling provides a rich plankton harvest, there is little fresh water runoff, and the water is warm and bright.

Atolls form when a volcanic island accumulates a skirt of coral around its shore, and then slowly subsiding at a rate equal to the growth rate of the coral. The central volcanic island would eventually sink from view, but the coral could grow continuously atop skeletons of past generations to maintain a living presence near the surface. Note in Figure 12.27 that the island begins with a fringing reef, passes through a barrier reef stage as it sinks, and eventually becomes an atoll as the peak disappears beneath the ocean surface. Should the island subside faster than about 1 centimeter per year (the growth rate of coral), all trace of both the island and the reefs would disappear. The submerged island might become a guyot.

4. How do human activities interfere with coastal processes? What steps can be taken to minimize loss of life and property along U.S. coasts?

Beaches exist in a tenuous balance between accumulation and destruction, and human activity can tip the balance one way or the other. We often divert rivers, build harbors, and develop property with surprisingly little understanding of the impact our actions will have on the adjacent coast. Residents of erosional coasts can only accept the inexorable loss of their property to the attack of natural forces, but residents of depositional coasts are sometimes presented with alternatives. The choices are almost never simple. For example, should rivers be dammed to control devastating floods? If the dams are built, they will trap sediments on their way from mountains to coast. Beaches within the coastal cell fed by the dammed river will shrink because the sand on which they depend to replenish losses at the shore is blocked. Alarmed coastal residents will then take steps to hang onto whatever sand remains. They may try to trap "their" beaches by erecting groins to stop the longshore transport of sediments. This temporary expedient usually accelerates erosion downcoast. Diminished beaches then expose shore cliffs to accelerated erosion -- wind wave energy that would have harmlessly churned sand grains now speeds the destruction of natural and artificial structures. Was protection from periodic flooding worth the loss of the beach? I suppose it depends on where your property is situated!

Shores that look permanent through the short perspective of a human lifetime are in fact among the most ephemeral of all marine structures. The only way to prevent the loss of life and property is not to build close to shore.

Thinking Analytically

1. Hurricane Isabel struck the east coast of the United States in September, 2003. Consult contemporaneous accounts and estimate the relative importance of high wind, heavy rain (and attendant flooding), and storm surge in the final damage toll.

You're on your own here. A search engine like Google © or Yahoo © might be a useful place to begin.

2. Figure 12.1 shows the hypothetical result of a rise in sea level of 60 meters (200 feet) in Florida. Obtain an elevation map of your favorite coastal town and see where the coastline would be if sea level rose 60 meters (200 feet) there.

I live in a coastal town in southern California that is home to the largest private yacht harbor in the world. Much of the town is built on a barrier and a series of dredge-filled islands "reclaimed" from a large bay. A 60 meter rise in sea level would take out the whole city and reach as far inland as Disneyland. Even a modest rise of 1 meter (3.3 feet) would do immeasurable damage to infrastructure and property in my town. Yours?

3. *About 5,000 cubic meters of sand is eroding every year from the beach shown in Figure 12.14b and d. Find the weight limit for dump trucks in your state. Now find the weight of a cubic meter of sand. How many dump truck loads would be required to make up this deficit?*

As a rule of thumb, sand weighs about 2 grams per cubic centimeter. (Granite has a mass of about 2.6 grams per cubic centimeter, but sand is neither pure granite nor a solid.) A cubic meter of sand would weigh about 2 million grams, or 2.2 tons. Check your State's public transportation law website and take it from there.

CHAPTER 13

LIFE IN THE OCEAN

Reviewing What You've Learned

1. What is the ultimate source of the energy used by most living things?

Nearly all the energy marine organisms need to function comes directly or indirectly from the sun. Light energy from the sun is trapped by chlorophyll in organisms called producers (certain bacteria, algae, and green plants) and changed into chemical energy. The chemical energy is used to build simple carbohydrates and other organic molecules -- food -- which is then used by the plant or eaten by animals (or other organisms) called consumers.

Another method of binding energy into carbohydrates is chemosynthesis, employed by a few relatively simple forms of life. Chemosynthetic organisms produce usable energy directly from energy-rich inorganic molecules available in the environment rather than from the sun. Chemosynthetic activity predominates in the deep ocean, particularly at the hydrothermal vents at tectonic spreading centers.

2. What do primary producers produce? How is productivity expressed?

Primary productivity is the synthesis of organic materials from inorganic substances by photosynthesis or chemosynthesis. The organic material produced is usually glucose. The source of carbon for glucose is dissolved CO_2. Primary productivity is expressed in grams of carbon bound into organic material per square meter of ocean surface area per year (g $C/m^2/yr$).

3. What is an autotroph? A heterotroph? How are they similar? How are they different?

Photosynthetic and chemosynthetic organisms can be called either primary producers or autotrophs because they make their own food. The bodies of autotrophs are rich sources of chemical energy for any organisms capable of consuming them. Heterotrophs are organisms such as animals that must consume other organisms because they are unable to synthesize their own food molecules. Some heterotrophs consume autotrophs, and some consume other heterotrophs. Plants are autotrophs; animals are heterotrophs.

95

4. What is a trophic pyramid? What is the relationship of organisms in a trophic pyramid. Does this have anything to do with food webs?

A trophic pyramid is a feeding hierarchy -- a construct showing the "who eats whom" relationships within a community. The primary producers shown at the bottom of the pyramid in Figure 13.5 are chlorophyll-containing photosynthesizers. The animal heterotrophs that eat them are called primary consumers (or herbivores), the animals that eat them are called secondary consumers, and so on to the top consumer (or top carnivore).

A trophic pyramid implies an oversimplistic view of a marine community. Real communities are more accurately described as food webs, an example of which is provided as Figure 13.6. A food web is a group of organisms linked by complex feeding relationships in which the flow of energy can be followed from primary producers through consumers. Organisms in a food web almost always have some choices of food species.

5. Briefly describe the operation of the carbon and nitrogen cycles.

The largest of all biogeochemical cycles is the global <u>carbon</u> cycle (Figure 13.8). Because of its ability to form long chains to which other atoms can attach, carbon is considered the basic building block of all life on Earth. Carbon enters the atmosphere by the respiration of living organisms (as carbon dioxide), volcanic eruptions that release carbon from rocks deep in Earth's crust, the burning of fossil fuels, and other sources. Large and small plants (and plant-like organisms) capture sunlight, and use this energy to incorporate, or *fix*, CO_2 into organic molecules. Some of these molecules are used as food, and some as structural components. When an animal eats a plant (or plant-like organism), about 45% of the ingested carbon is used for growth. The end product of respiration is CO_2, a gas eventually lost to the atmosphere. Plants use this CO_2 as a carbon source, and the cycle begins gain.

<u>Nitrogen</u> is a critical component of proteins, chlorophyll, and nucleic acids. Like carbon, nitrogen may be found in the bodies of organisms, as a dissolved gas (N_2), and as dissolved organic matter. Most organisms cannot use the free nitrogen in the atmosphere and ocean directly. It must first be bound with oxygen or hydrogen, or *fixed*, into usable chemical forms by specialized organisms, usually bacteria or cyanobacteria. After being assimilated by small plants and plantlike organisms, nitrogen is recycled as animals consume them, then excrete primarily ammonium and urea. These reduced forms of nitrogen are then oxidized back into nitrate, via nitrite, by nitrifying bacteria. The cycle begins again.

6. Name and briefly discuss five physical factors of the marine environment that impact living organisms. How is each different in the ocean from the land?

Transparency is important because photosynthesizers require light to make food. The depth to which light penetrates is limited by the number and characteristics of particles in the water. These particles, which may include suspended sediments, dust-like bits of once-living tissue, or the organisms themselves, scatter and absorb light. Most of the biological productivity of the ocean occurs in an area near the surface called the euphotic zone. This is where marine plants trap more energy than they use. Though it is difficult to generalize for the ocean as a whole, the euphotic zone typically extends to a depth of approximately 40 meters (130 feet) in mid-latitudes. On land, photosynthesis proceeds at or just above ground level. Light cannot penetrate soil, so plants cannot photosynthesize below the soil surface as they can below the ocean's surface. Also, land plants require strong structures (limbs, trunks) to support themselves in the sunlit air, whereas marine plants can depend on the buoyancy of water to maintain their position. Marine plants would seem to have the better situation.

Temperature determines an organism's metabolic rate, the rate at which energy-releasing reactions proceed. Too high a temperature and the organism cooks, of course, but in general, the warmer the temperature, the greater the organism's metabolic rate. Ocean temperature varies with depth and latitude. The average temperature of the world ocean is only a few degrees above freezing, with warmer water found only in the lighted surface zones of the temperate and tropical ocean, and in rare, deep, warm chemosynthetic communities. Though temperature ranges of the ocean are considerable, they are much narrower than comparable ranges on land. Marine organisms have the definite advantage: they are almost never exposed to sustained temperatures above 30°C (86°F), while some terrestrial species must tolerate long periods when temperature reaches or even exceeds 60°C (140°F).

Dissolved nutrients are compounds that organisms require for the production of organic matter, for structural parts, and for the manipulation of energy. A few of these necessary nutrients are always present in seawater, but most are not readily available. The main inorganic nutrients required in primary productivity include nitrogen (as nitrate, NO_3^-) and phosphorus (as phosphate, PO_4^{3-}). As any gardener knows, land plants require fertilizer -- mainly nitrates and phosphates -- for success. Ocean gardeners would have more trouble raising crops than their terrestrial counterparts, though, because the most fertile ocean water contains only about 1/10,000 the available nitrogen of topsoil. Phosphorus is even scarcer in the ocean, but fortunately less of it is required by living things. Nitrogen and phosphorus are often depleted by autotrophs during times of high productivity and rapid reproduction. Though primary productivity may be very high when light is available, the total mass of living material cannot increase until more inorganic nutrients are made available by recycling, upwelling, runoff from land, or other means.

Dissolved gases are required by all living organisms, oceanic or terrestrial. We land organisms dissolve gases in a fluid layer on the lining or our lungs, but marine organisms make use of gases already dissolved in the matrix in which they live. Rapid photosynthesis at the surface lowers CO_2 concentrations and increases the quantity of dissolved oxygen. Oxygen is least plentiful just below the limit of photosynthesis because of respiration by many small animals at middle depths. Low oxygen levels can sometimes be a problem at the ocean surface. Plants produce more oxygen than they use, but they produce it only during daylight hours. The continuing respiration of plants at night will sometimes remove much of the oxygen from the surrounding water. In extreme cases this oxygen depletion may lead to the death of the plants and animals in the area, a phenomenon most noticeable in enclosed coastal waters during spring and fall plankton blooms. In general, inappropriate quantities of dissolved gases are more of a problem for marine organisms than for terrestrial ones.

Hydrostatic pressure, the pressure caused by the great weight of water above a marine organism, usually presents very little difficulty unless the organism swims rapidly up or down in the water column. In fact, the situation in the ocean is parallel to that on land. Land animals live in air pressurized by the weight of the atmosphere above them (1 kilogram per square centimeter, or 14.7 pound per square inch, at sea level) without experiencing any problems. In terrestrial organisms, the pressure inside an organism and outside it is virtually the same. In marine organisms the pressure is usually much higher, but again, there is little difference between the pressure within the organism and outside it.

7. How is the marine environment classified? Which scheme is most useful? Justify your answer.

Scientists divide the marine environment into zones, areas with homogeneous physical features. Zonation may be based on light, depth, salinity, or water mass composition. The scheme that is most useful depends on the particular factor (or set of factors) in which you are interested. A researcher studying primary productivity might find zonation by illumination to be the most useful division (photic zone, euphotic zone, aphotic zone, etc.), while someone studying organisms' responses to sediment size would prefer benthic division based on position along the continental margin (bathyal zone, abyssal zone, etc.). One's perception of zonation depends on one's interests!

8. Would you support expenditure of government funds to search for asteroids or other bodies on a collision course with Earth? What would the public's response be to discovery of a serious threat?

The near-space environment is being scanned for Earth-crossing comets and asteroids (bodies whose orbits intersect that of the Earth). The resources dedicated to this task are meager, however, and it will be decades before most of the potential threats are catalogued. Congress has not supported a significant increase in funds for this purpose, and the attention of the public waxes and wanes with Hollywood's interest in the topic.

Some indication of public response can be gleaned from the same Hollywood movies. Among the earliest and best of these is a 1950s George Pal production titled "When Worlds Collide" that greatly influenced me to study science when I was a small and impressionable boy.

Thinking Critically

1. The second law of thermodynamics states that entropy (disorganization) tends to increase with time. Living things tend to become more complex with time (embryos grow to adults, populations evolve). How can that be?

Living organisms are complex assemblies; they need energy to build that complexity from large numbers of simple molecules. The second law of thermodynamics shows, however, that disorder inevitably tends to increase in the universe as time passes. Things run down, become disorganized, and break. Entropy is a measure of this disorder.

Living things are not exempt from the second law of thermodynamics, but they can delay that inevitable descent into entropy because the transformation of energy in living things allows a temporary and local remission of the second law. Thus, living organisms might be defined as localized regions where the flow of energy results in *increased* order; that is, areas of great complexity and low entropy. Car engines use the energy in gasoline only to move. Organisms use energy in food to move, to maintain their highly complex organization, and to grow. Cars and computers don't fix themselves; fish and people do. This sophisticated use of energy seems to be a basic attribute of living things.

The main source of energy for living things on Earth is the sun. Life prospers, becomes ever more complex, and evolves into millions of forms by accepting sunlight and radiating waste heat to the cold of space. With only a few exceptions, organisms get their power directly or indirectly through the capture, storage, and transmission of energy from sunlight. Light energy is transformed into chemical energy and finally into heat as organisms temporarily forestall the disorderly fate decreed by the second law.

In the long run, however, this fate is inescapable. The sun powers the biological systems of Earth, and the sun is "winding down." As the sun dies, so will life on Earth. Death is marked by an irreversible increase in the entropy of an organism. Without an unbroken flow of energy, the complexity of living things returns to the simplicity of atoms and heat. The second law always wins in the end.

2. Can you suggest any ways humans might be altering biogeochemical cycles?

Perhaps the most obvious alteration concerns the accelerated release into the atmosphere of carbon dioxide through the burning of fossil fuels. Evidence strongly suggests the advent of human-induced global warming (news of which awaits you in Chapter 18). Subtle changes in temperature can greatly affect the success and distribution of organisms that, in turn, can affect nutrient cycles, pH, parasite loads, and tens of other physical and biological environmental factors.

On a smaller scale, local introduction into the environment of heavy metals, or acids (or bases), or nutrients (such as phosphates or iron) can affect the uptake and release of other elements or compounds. Such substances can become limiting under the wrong conditions, and damage populations or communities.

3. What is a limiting factor? Can you think of some examples not given in the text?

Too much or too little of a single physical factor can adversely affect the function of an organism. We call that factor a limiting factor, a physical or biological necessity whose presence in inappropriate amounts limits the normal action of the organism.

Imagine a situation in which a radiolarian, a small planktonic protist equipped with a beautiful silica-rich shell, finds itself in a water mass almost completely devoid of silicate minerals. Although conditions for its growth and development may otherwise be ideal, the radiolarian will not thrive because its shell cannot grow. It is limited by the lack of silicate minerals in the surrounding water.

4. How is evolution by natural selection thought to work?

Here are the main points:

In any group of organisms, more offspring are produced than can survive to reproductive age.

Random variations occur in all organisms. Some of these variations are inheritable; that is, they can be passed on to the offspring.

100

Some inheritable traits increase the probability that the organisms possessing them will survive. These are favorable traits.

Because bearers of favorable traits are more likely to survive, they are also more likely to reproduce successfully than bearers of unfavorable traits. Thus, favorable traits tend to accumulate in the population; they are *selected*. Note this is not a random process.

The physical and biological (*natural*) environment itself does the selection. Favorable traits are retained because they contribute to the organism's success in its environment. These traits show up more often in succeeding generations if the environment stays the same. If the environment changes, other traits become favorable and the organisms with those traits live most effectively in the environment.

5. How would you define biological success? Does success depend on the size of an organism? Its beauty? What amount of space it controls? Its numbers?

Success in biology depends on only one thing: *offspring numbers*. Practically by definition, a successful organism will leave behind a great many descendants, each of which (if successful) will do the same.

6. How does a natural system of classification differ from an artificial system? Can you give an example of each? Was the hierarchy-based system invented by Linnaeus natural or artificial? What is a hierarchy-based system?

An artificial system of classification is a system of classifying animals based on their exterior similarities. An example of an artificial system of classification is the arrangement of compact disks by jacket color, or manufacturer, or label design. By contrast, the natural system of classification for living organisms (such as biologist use today) relies on an organism's structural and biochemical similarities. We place all insects together regardless of their flying ability just as we place all books by Melville together, all compositions of Haydn together, and all sea stars together because each group has a common underlying natural origin. The groups are arranged systematically -- that is, in some order that makes structural and evolutionary sense. In compact disks, it makes sense to arrange them by the type of music recorded on them, the reason you bought them in the first place.

Linnaeus's system of classification was decidedly natural. Though Darwin's insights into evolutionary relationships were nearly a century in the future, Linnaeus's understanding of the relationships between organisms, and his ability to arrange organisms into like categories, was remarkable. His was a system of classification based on hierarchy, a grouping of objects by degrees of complexity, grade, or class. In this boxes-within-boxes approach, sets of small categories are nested within larger categories. Linnaeus devised names for the categories, starting with kingdom (the largest category) and passing down through phylum, class, order, family, and genus, to species (the smallest category).

Thinking Analytically

1. An apple contains about 50 grams of carbon. How many "apple equivalents" would typically be produced each year in a square meter of open ocean surface area? In a kelp bed? (Hint: See Figure 13.3.)

As Figure 13.3 suggests, average productivity for the open ocean is around $100gC/m^2/yr$. This would yield the equivalent of perhaps two apples' worth of fixed carbon annually (which, when you visualize it, is quite impressive). A healthy kelp bed could sustain much higher productivity, perhaps as much as $1500gC/m^2/yr$, or 30 apples' worth, per year. Pictured this way, oceanic productivity can be seen as extraordinarily high.

2. Researchers believe there may be as many as 100 million species of living things on and in Earth. We know of about a million species so far. Where do you think the rest of the species are hiding out?

Most of the new species will surely be very small – there must be very few new warthogs or sharks left to find. Marine researchers combing through marine sediments, sand in tropical reefs, and pores in rocks have discovered an astonishing array of interstitial fauna –organisms literally living "between the spaces." Terrestrial scientists seem to have a seemingly unending array of new insects to discover, but only one genus of marine insect (*Halobatis*) has been found, so isn't going to prove a rich trove. As the years pass, our ability to detect and study new forms of life will make the list of known, named species much longer.

Chapter 14

Plankton, Algae, and Plants

Reviewing What You've Learned

1. What do primary producers produce? How is productivity expressed?

The synthesis of organic materials from inorganic substances by photosynthesis or chemosynthesis is called primary productivity. Primary productivity is expressed in grams of carbon bound into organic material per square meter of ocean surface area per year ($gC/m^2/yr$). The organic material produced is usually glucose, a carbohydrate. The source of carbon for glucose is dissolved CO_2.

2. What is plankton? How is plankton collected? How are members of the plankton community differentiated? How is zooplankton different from phytoplankton?

Plankton drift or swim weakly, going where the ocean goes, unable to move consistently against waves or current flow. The plankton contains many different plant-like species and virtually every major group of animals. The term is not a collective natural category like mollusks or algae, which would imply an ancestral relationship between the organisms; instead it describes a basic ecological connection. Members of the plankton community, informally referred to as plankters, can and do interact with one another -- there is grazing, predation, parasitism, and competition among members of this dynamic group.

Plankton are usually collected in fine-mesh nets customarily made of nylon or Dacron cloth woven in a fine interlocking pattern to assure consistent spacing between threads. The net is hauled slowly for a known distance behind a ship, or cast to a set depth, and then reeled in. Trapped organisms are flushed to the net's pointed end and carefully removed for analysis. Quantitative analysis of plankton requires the organisms and an estimate of the sampled volume of water. Very small plankton can slip through a plankton net. Their capture and study requires concentration by centrifuge, or entrapment by a plankton filter through which water is drawn. The filter is later disassembled and the plankton studied in place.

Autotrophic plankton is generally called phytoplankton, a term derived from the Greek word meaning plant. A huge, nearly invisible mass of phytoplankton drifts within the sunlight surface layer of the world ocean. Phytoplankton is critical to all life on Earth because of its great contribution to food webs and its generation of large amounts of atmospheric oxygen through photosynthesis. Heterotrophic plankton -- the planktonic organisms that eat these primary producers -- is collectively called zooplankton. Zooplankters are the most numerous primary consumers of the ocean. They graze on the diatoms, dinoflagellates, and other phytoplankton at the bottom of the trophic pyramid in a way analogous to cows grazing on grass. The variety of zooplankton is surprising; nearly every major animal group is represented.

3. Describe the most abundant and important phytoplankters. Which are most efficient in converting solar energy to energy in chemical bonds?

The dominant photosynthetic organisms in the plankton -- and in the world -- are the <u>diatoms</u>. More than 5,600 species of diatoms are known to exist. The larger species are barely visible to the unaided eye. Most are round, but some are elongated or branched or triangular. Typical diatoms are shown in Figure 14.3. Diatoms are encased in a "shell" (actually a frustule) that consists of silica (SiO_2), giving this beautiful covering the optical, physical, and chemical characteristics of glass -- an ideal protective window for a photosynthesizer. Inside the frustule lies the most nearly perfect photosynthetic machine yet to evolve on the planet. Fully 55% of the energy of sunlight absorbed by a diatom can be converted into the energy of carbohydrate chemical bonds, one of the most efficient energy conversion rates known. Excess oxygen not needed in the cell's respiration is released through the perforations in the frustule into the water. Marine animals absorb some oxygen, some is incorporated into bottom sediments, and some diffuses into the atmosphere. Most of the oxygen we breathe has moved recently through the many glistening pores of diatoms.

<u>Dinoflagellates</u> are the second most important phytoplankters. Like diatoms, they are are single-celled autotrophs. A few species live within the tissues of other organisms, but the great majority of dinoflagellates live free in the water. Most have two whip-like projections called flagella in channels grooved in their protective outer covering of cellulose (see Figure 14.5). One flagellum drives the organism forward while the other causes it to rotate in the water. Their flagella allow dinoflagellates to adjust orientation and vertical position to make the best photosynthetic use of available light. Dinoflagellates are responsible for red tides.

<u>Coccolithophores</u>. Most other types of phytoplankton are extraordinarily small, and so are called nanoplankton. The coccolithophores, for example, are tiny single cells covered with discs of calcium carbonate (coccoliths) fixed to the outside of their cell walls (see Figure 14.6). Coccolithophores live near the ocean surface in brightly lighted areas. Coccoliths can build seabed deposits of ooze. The famous White Cliffs of Dover in England consist largely of fossil coccolith ooze deposits uplifted by geological forces.

Each of these autotrophs fixes carbon into glucose by photosynthesis.

4. How can primary productivity be measured? Which method is considered the most accurate?

Since we know the formula for photosynthesis, measuring any one component of photosynthesis will tell us about the others. For example, from a measurement of oxygen production by primary producers we can calculate the rate of carbohydrate production. One way to do this is to place identical water samples, containing identical populations of organisms, into pairs of identical transparent and opaque bottles, and then suspend chains of the bottles from the side of a ship from dawn until noon. The transparent bottles admit light; the opaque bottles block it. Autotrophs within the transparent bottles produce and consume oxygen; autotrophs within the opaque bottles consume oxygen but are unable to generate it; and heterotrophs consume oxygen in both bottles. The difference in the quantity of oxygen in each pair of bottles over time gives an indication of the net rate of productivity at each depth.

A more sensitive way to measure productivity is by tagging atoms of carbon in CO_2 with radioactive tracers. The rate at which the marked carbon is incorporated into carbohydrates is a direct indication of primary productivity. This procedure requires costly and delicate equipment but provides researchers with the most accurate estimates of productivity in controlled situations.

But ocean conditions are complex and variable, rarely controlled. A new and promising method of gauging productivity may be the most effective and useful of all. Recent advances in remote sensing have made it possible to estimate the chlorophyll content of ocean water from orbiting satellites (as in Figure 14.6). Because the amount of chlorophyll present is directly related to the rate of photosynthesis, chlorophyll content is a good indicator of productivity.

5. What are algae? Are all algae seaweeds? How are seaweeds classified? Which seaweeds live at the greatest depths? Why?

Algae is a collective term for autotrophs possessing chlorophyll and capable of photosynthesis but lacking vessels to conduct sap.

Not all algae are seaweeds: the single-celled diatoms and dinoflagellates discussed earlier are unicellular algae.

Seaweeds are classified by the presence of accessory pigments, colored compounds in their tissues. These accessory pigments (or masking pigments) are light absorbing compounds closely associated with chlorophyll molecules. Accessory pigments may be brown, tan, olive green, or red; they are what give most marine autotrophs, especially seaweeds, their characteristic color. Multicellular marine algae are segregated into three divisions based on their observable color. The green algae, with their unmasked chlorophyll, are the Chlorophyta, the brown algae Phaeophyta, and the red algae Rhodophyta. Phaeophytes are most familiar to beachcombers, and rhodophytes the most numerous in the ocean as a whole.

Rhodophytes can live in surprisingly <u>deep</u> water. They excel in dim light because their sophisticated accessory pigments absorb and transfer enough light energy to power photosynthetic activity at depths where human eyes cannot see light. The record depth for a photosynthesizer is held by a small rhodophyte discovered in 1984 at a depth of 268 meters (879 feet) on a previously undiscovered seamount in the clear tropical Caribbean.

6. Give examples of marine angiosperms. Are they vascular or nonvascular plants? Are they considered seaweeds?

Angiosperms are advanced vascular plants that reproduce with flowers and seeds. Most large land plants are angiosperms. Relatively few angiosperms live in water -- the advantages conferred by vessels and roots are largely unnecessary in an aquatic environment -- but a few species of angiosperms have recolonized the ocean. All of these have descended from land ancestors, and all live in shallow coastal water. Angiosperms live at the surface where the red light required for photosynthesis is abundant; they have no need for accessory pigments and their chlorophyll is unmasked. The most conspicuous marine angiosperms are the sea "grasses" (which are not true grasses), and the mangroves.

Many people lump the smaller marine angiosperms, especially the sea "grasses," in the informal seaweed group, but their resemblance to large marine algae is only superficial.

7. Of what commercial importance are marine plants?

The mucilaginous material that is so effective in making algal blades slick, lowering friction, and deterring grazers is harvested and made into an important commercial product called algin. When separated and purified, its long, intertwining molecules are used to stiffen fabrics, make adhesives, suspend water and oil together in salad dressings, prevent the formation of gritty crystals in ice cream, clarify beer and wine, and manufacture shoe stains, soaps, and shaving cream. Fast-food restaurants are using carrageen, a similar seaweed extract, to replace some of the fat in newly popular healthier hamburgers. These substances also prevent fire-extinguishing foams from dispersing, permit chocolate milk to remain on the refrigerator shelf without separating, and keep the abrasives in liquid car waxes from settling to the bottom of the bottle. In biological laboratories bacteria are cultured on agar made from seaweed extracts. Very likely even the ink that forms the letters you are now reading has an algin or carrageen component!

Seaweeds are eaten directly, too. Animals raised for food or fur eat the plant material remaining after the mucilaginous components are removed. Seaweeds' mineral content and roughage are also useful in human nutrition, and human consumption of seaweed is especially high in oriental countries -- the Japanese consume more than 150,000 metric tons of *nori* annually, primarily in the form of sushi. Seaweed and seaweed extracts are also eaten in Britain, Ireland, and New Zealand.

Seaweeds are big business. Algal products were key substances in the production of over $44 billion worth of products in 2003. The future may be even brighter. The complex biochemistry of the brown and red algae holds promise in pharmaceutical research, and seaweed-based drugs against parasitic infections, thyroid imbalance, kidney disease, high blood cholesterol, hypertension, and heavy metal poisoning have already been developed.

Thinking Critically

1. What factors limit productivity? What methods have marine producers evolved to cope with the lack of red light needed by chlorophyll for photosynthesis?

Photosynthetic autotrophs require four ingredients to produce carbohydrates: water, carbon dioxide, inorganic nutrients, and sunlight. Obviously, water is not a limiting factor in the ocean. Carbon dioxide is almost never a limiting factor either, because of its high solubility in water, and because of the large quantity of carbon dioxide dissolved in the ocean. So the two potential limiting factors in primary productivity are nutrients and light.

Lack of nutrients is the most common factor limiting primary productivity. Adding a few hundred kilograms of nitrate- and phosphate-rich lawn fertilizer to a calm, warm, sunlit ocean area depleted of nutrients will usually turn the place into a rich phytoplankton soup in a matter of hours or days.

If adequate nutrients are present, productivity depends on illumination. Too little light is obviously limiting for photosynthesizers; very little photosynthesis proceeds below 100 meters (330 feet), and no photosynthesizers are known to function below 268 meters (879 feet). Color is of critical importance. Chlorophyll is a green pigment and thus absorbs best in the red and violet wavelengths. Chlorophyll looks green because it reflects green light. But red light is effectively absorbed and converted into heat near the ocean surface; very little red light penetrates past three meters

(10 feet). Most marine autotrophs have evolved specialized accessory pigments that greatly enhance photosynthesis by absorbing the dim blue light at depth and transfer its energy to the adjcent chlorophyll molecules. Accessory pigments may be brown, tan, olive green, or red; they give most marine autotrophs, especially seaweeds, their characteristic color.

2. What is compensation depth? What happens to phytoplankton below that depth? To zooplankton?

Recall that autotrophs respire as they photosynthesize -- they use some of the carbohydrates and oxygen they produce. Carbohydrate production usually exceeds consumption, but not always. The deeper a phytoplankter's position, the less light it receives. At a certain depth, the production of carbohydrates and oxygen by photosynthesis through a day's time will exactly equal the consumption of carbohydrates and oxygen by respiration. This "break even" depth is called the <u>compensation depth</u>. Compensation depth usually corresponds to the depth to which about 1% of surface light penetrates.

Zooplankters are animals, not primary producers; the concept of compensation depth for a non-autotroph is meaningless. By the way, this makes a *nice* test question!

3. Where in the ocean is plankton productivity the greatest? Why?

Because of coastal upwelling and land runoff, nutrient levels are highest near the continents. Plankton is most abundant there, and productivity highest. The water above some continental shelves sustains productivity in excess of 1 $gC/m^2/day$! But what of the open ocean? Where is productivity greatest away from land?

The open tropical oceans have abundant sunlight and CO_2 but are generally deficient in surface nutrients because the strong thermocline discourages the vertical mixing necessary to bring nutrients from the lower depths. The tropical oceans away from land are therefore oceanic deserts nearly devoid of visible plankton. The typical clarity of tropical water underscores this point. In most of the tropics productivity rarely exceeds 30 $gC/m^2/yr$, and seasonal fluctuation in productivity is low.

At very high latitudes, the low sun angle, reduced light penetration due to ice cover, and weeks or months of darkness in winter severely limit productivity. At the height of summer, however, 24-hour daylight, a lack of surface ice, and the presence of upwelled nutrients can lead to spectacular plankton blooms. The surface of some sheltered bays can look like tomato soup because dinoflagellates and other plankton are so abundant. This bloom cannot last because nutrients are not quickly recycled and because the sun is above the critical angle for a few weeks at best. The short-lived summer peak does not compensate for the long, unproductive winter months.

With the tropics generally out of the running for reasons of nutrient deficiency and the north polar ocean suffering from slow nutrient turnover and low illumination, the overall productivity prize is left to the temperate and southern subpolar zones. Thanks to the dependable light and moderate nutrient supply, annual production in the nearshore temperate and southern subpolar ocean areas is the greatest of any open ocean area. Typical productivity in the temperate zone is about 120 $gC/m^2/yr$. In ideal conditions southern subpolar productivity can approach 250 $gC/m^2/yr$!

4. How does a nonvascular plant differ from a vascular plant? Why are most marine plants nonvascular?

Nearly all land plants larger than a thumbtack have such conducting vessels and thus are known as <u>vascular</u> plants. In land plants two of the four ingredients needed for photosynthesis are in the air (carbon dioxide and sunlight) and two are in the soil (water and inorganic nutrients). But for photosynthesis to proceed all four must be present at the same place -- in the leaves. A maple tree absorbs water and nutrients from the soil and its vessels transport them to the leaves.

Algae are <u>nonvascular</u> plants; that is, they do not have vessels. Algae require the same four ingredients for photosynthesis as vascular plants, but in their case the ingredients are simultaneously present in one location. Either the alga is very small (perhaps a single cell) and lives in a moist spot on land where conditions are ideal, or it lives in water.

Thinking Analytically

1. Is phytoplankton productivity highest at the ocean surface? What advantage would optimum productivity at a depth below the surface provide to phytoplankton?

As can be seen from the curve in Figure 14.8 on page 343, the depth of greatest productivity is (in this example) perhaps 20 meters below the surface. Brighter light nearer the ocean surface inhibits photosynthesis by overwhelming the producers' photochemical systems. But imagine for a moment that photosynthesis were optimized for surface illumination conditions – that the depth of greatest productivity is zero meters. Net productivity is the green area of the curve. Can you see how much <u>smaller</u> the green area would be if the "bulge" were moved to the surface? Phytoplankters are most productive at a depth below the surface because their biochemical machinery can benefit from the upper and lower "tail" of the productivity curve.

2. *Imagine a small tide-washed inlet on the coast of western Canada about the size of an Olympic swimming pool. Given optimal nutrients, a stable substrate, and the sunlight of summer, estimate approximately how many kilograms (wet weight) of seaweed as productive as Postelia could be produced in the inlet in one month. How does this compare with a field of alfalfa of the same surface area? Begin with a review of Figure 13.3 and the biology of seaweeds in this chapter.*

An Olympic swimming pool measures 50 meters by 25 meters, a surface area of 1,250 square meters. As you read on page 352, the seaweed *Postelia* – under optimal conditions – can bind an astonishing 14,600 $gC/m^2/yr$, the highest primary productivity recorded for any autotroph. We want to know monthly productivity, so divide 14,600 $gC/m^2/yr$ by 12 to obtain 1,217 $gC/m^2/yr$. Multiply by the area of the swimming pool, and we find a startling productivity of about 1.5 million grams per month, or about 3,350 pounds (1.68 tons). An alfalfa field is only about 1/10th as productive (see page 306). Not all glucose produced is turned into biomass – much of it undergoes respiration to produce energy.

CHAPTER 15

MARINE ANIMALS

Reviewing What You've Learned

1. What is an animal? How is an animal different from an autotroph?

Animals are active multicellular heterotrophs. They cannot synthesize their own food and must ultimately depend on primary producers (autotrophs) for nutrition. Years ago, protozoa (like amoeba or paramecium) were called "single celled animals," but animals are now, by definition, multicellular. The protozoa are termed "animal-like."

2. Which animal phylum is most successful? How is success defined? What structural advances contribute most to that phylum's immense success?

The phylum Arthropoda, a group that includes the lobsters, shrimp, crabs, krill, and barnacles, is a phylum of superlatives. Over a million species of arthropods are now known, but some experts suggest more than 10 million species of insects -- the most numerous members of this phylum -- may exist on Earth! Considering their numbers, Arthropods are the most successful of Earth's animal phyla, occupying the greatest variety of habitats, consuming the greatest quantities of food, and existing in almost unimaginable numbers.

Arthropods exhibit three remarkable evolutionary advances that have led to their great success:

An *exoskeleton*, a strong, lightweight, form-fitted external covering and support.

Striated muscle, a quick, strong, lightweight form of muscle that makes rapid movement and flight possible.

Articulation the ability to bend appendages at specific points. The appendages of more primitive phyla can usually bend anywhere (like wet spaghetti), but arthropod appendages bend at a joint. There are no ball-and-socket joints in arthropods; instead each joint along an appendage moves through a different plane to ensure a full range of motion.

Most important of these advances is the exoskeleton. Unlike the often-cumbersome shell of a gastropod, the exoskeleton of an arthropod fits and articulates like a finely tailored suit of armor. It is made in part of a tough nitrogen-rich carbohydrate called chitin that may be strengthened by calcium carbonate. Its three layers serve to waterproof the covering, tint it a protective color, and make it resilient and strong. Muscles within the animal are attached to the exoskeleton to move the appendages.

3. Which phylum contains the largest representatives? The most intelligent?

Cephalopod mollusks are the largest and most intelligent invertebrates. These highly evolved molluskan predators include the nautiluses, octopuses, and squid. These well-named animals have a head surrounded by a foot divided into tentacles. Cephalopods can move by creeping across the bottom, by swimming with special fins, or by squirting jets of water from an interior cavity.

Squids can grow to surprising sizes (see again Figure 15.15) -- the record length, including tentacles, is 18 meters (59 feet)! Most are much smaller.

Octopuses are the most intelligent mollusks. About as smart as puppies and with even better eyesight, some nearshore species of octopuses kept in captivity soon learn to recognize their keepers and forage at night through adjacent aquariums for tidbits. Octopuses use their visual acuity and intelligence to good advantage in their intertidal or subtidal homes, memorizing the positions of hiding places, escape holes, and good hunting locations.

Only about 450 species of cephalopods live today, a small percentage of the number of species known from fossils. Their small number of species suggests that sophistication is no guarantee of biological success.

4. Which phyla exhibit radial symmetry? Bilateral symmetry? No symmetry?

In radial symmetry, body parts radiate from a central axis like the spokes of a wheel. Cnidarians and echinoderms are examples.

Bilaterally symmetrical animals have a left side and a right side that are mirror images of each other. The worm phyla, mollusks, arthropods, and chordates are examples.

Leucon sponges are occasionally asymmetrical, growing in large blobs in seemingly random directions.

5. What are the classes of living fishes? Which is considered most primitive? Most advanced? Which class has the largest individuals? Which is the most economically important?

Fishes are divided into three major groups: The primitive jawless snake-like fishes of the Class <u>Agnatha</u>, the cartilaginous fishes of the Class <u>Chondrichthyes</u>, and the advanced bony fishes of the Class <u>Osteichthyes</u>.

Sharks and rays in the class Chondrichthyes tend to be <u>larger</u> than bony fishes, and except for some whales, sharks are the largest living vertebrates. The bony fishes, members of Class Osteichthyes, owe much of their great success to the hard, strong, lightweight skeleton that supports them. These most numerous of fish -- and most numerous, most diverse, and successful of all vertebrates -- are found in almost every marine habitat from tidepools to the abyssal depths. Their numbers include the air breathing lungfishes and lobe-finned coelacanths, whose ancient relatives broke from the path of fish evolution to establish the dynasties of land vertebrates.

Bony fishes (Osteichthyes) are the most <u>economically important</u> marine organisms.

6. How do fishes swim? What problems are associated with swimming? How do fishes overcome these problems? How do fishes make defensive use of color, shape, and schooling behavior?

A fish's forward thrust comes from the combined effort of body and fins. Muscles within slender flexible fish (such as eels) cause the body to undulate in S-shaped waves that pass down the body from head to tail in a snake-like motion. The eel pushes forward against the water much as a snake pushes against the ground. This type of movement is not very efficient, however; the body must wave back and forth across a considerable distance exposing a large frontal area to the water, and the long body length (relative to width) needed to propagate the wave requires increased surface area which increases drag. More advanced fishes have a relatively inflexible body that undulates rapidly through a shorter distance; they also have a hinged, scythe-like tail to couple muscular energy to the water. The fish's body can be shorter and can face more squarely in the direction of travel; the drag losses are lower.

Active fish usually have streamlined shapes that make their propulsive efforts more effective. A fish's resistance to movement, or drag, is determined by frontal area, body contour, and surface texture. Drag increases geometrically with increasing speed. Faster-swimming fish are therefore more highly modified to minimize the slowing effects of the dense, relatively sticky medium in which they live.

Defense measures are well advanced in fishes. Some fishes such as sea horses (and their relatives, the box fishes) depend on the simple expedient of armor plating for protection. Others, such as the puffer fish, inflate with water and erect bristly spines to become less attractive as a snack. More subtle means of offense or defense depend on trickery -- looking like something you're not, or changing color to blend with the background. These kinds of cryptic coloration or <u>camouflage</u> may be active or passive. Actively swimming fish employ a different method of color blending called countershading. Their dark tops and silvery bottoms make the fish less obvious to predators above or below.

About a quarter of all bony fish species exhibit <u>schooling</u> behavior at some time during their life cycle. A fish school is a massed group of individuals of a single species and size class packed closely together and moving as a unit. There is no leadership in fish schools, and the movement of fish within them seems to be controlled automatically by direct interaction between lateral-line sensors and the locomotor muscles themselves.

7. How do fishes "breathe" underwater? Do marine fishes experience any unusual problems by being immersed in seawater?

Gas exchange, the process of bringing oxygen into the body and eliminating carbon dioxide, is essential to all animals. At first glance the task may seem more difficult for water breathers than for air breathers, but air-breathing animals add an extra step. We air breathers must first dissolve gases in a thin film of water in our lungs before they can diffuse across a membrane.

Fish take in water containing dissolved oxygen at the mouth, pump it past fine gill membranes, and exhaust it through rear-facing gill slits. The higher concentration of free oxygen dissolved in the water causes oxygen to diffuse through the gill membranes into the animal; the higher concentration of CO_2 dissolved in the blood causes CO_2 to diffuse through the gill membranes to the outside. The gill membranes themselves are arranged in thin filaments and plates efficiently packaged into a very small space (Figure 15.32). Water and blood circulate in opposite directions -- in a countercurrent flow -- that increases transfer efficiency.

An active fish like a mackerel requires so much oxygen and generates so much waste CO_2 that its gill surface area must be ten times its body surface area. (Sedentary fishes have proportionally less gill area.) With their large gill area and countercurrent flow, active fish extract about 85% of the dissolved oxygen in water flowing past their gills. Air breathing vertebrates, by contrast, extract only about 25% of the oxygen from air entering their lungs.

Living in seawater can cause osmoregulatory difficulties. Marine teleosts (bony fishes), with their higher relative internal concentration of water (and lower concentration of salts) per unit of fluid volume, continuously *lose* water to their environment; fresh water teleosts, with their lower relative internal concentration of water, constantly *absorb* water from their environment. If these fishes were incapable of osmoregulation -- if they had no active way of adjusting their internal salt concentration -- they would quickly die of fluid imbalance.

114

Figure 15.33 summarizes the ways bony fishes cope with osmoregulatory difficulties. The skin of both fresh water and marine teleosts is nearly impermeable to water and salts. A fresh water fish does not drink water, uses large kidneys to generate copious quantities of dilute urine to export the invading water, and actively absorbs salts through its gills from the surrounding water and from its urine. A marine fish makes only small quantities of urine, actively drinks seawater (some species drinking up to 25% of their body weight per day), and eliminates excess salts through special salt secreting cells in the gills. Bony fish consume a substantial amount of energy in these essential osmoregulatory tasks.

8. How are seabirds different from land birds? Are there more species of seabirds than land birds?

Truly marine birds are typically fairly large animals -- there are no sparrow-sized pelagic seabirds. If they fly, they have very long, thin, pointed, cupped wings. (Penguins, which "fly" through water, have a small wing hydrodynamically adapted to a different fluid than air.) The legs of marine birds are usually farther back on the body than the legs of a terrestrial bird, so they stand up straighter than, say, a chicken or pigeon. The eggs of seabirds tend to be more sharply conical, presumably so they'll roll in a smaller diameter circle if they fall out of the nest, not a bad adaptation for an animal whose nests are often built in confined areas along the edges of cliffs. Marine birds often have salt glands in their heads, tissue able to extract the excess salt from the blood that results from drinking seawater. Their sense of smell is usually highly developed. Marine birds have exceedingly lightweight skeletons and other weight-saving adaptations that enable them to spend very long times aloft at sea with relatively little energy output. Many, like the albatrosses, also have highly developed homing systems that enable them to return to the nesting site after months or years at sea.

Only about 3% of all known bird species (270) qualify as seabirds -- adaptation to the severe marine environment is perhaps more difficult to achieve than to the land environment.

9. What characteristics are shared by all marine mammals?

All marine mammals share four common features:

They have a streamlined body shape with limbs adapted for swimming that makes an aquatic lifestyle possible. Thin, stiff flippers and tail flukes situated at the rear of the animal drive it forward, and similarly shaped forelimbs act as rudders for directional control. Drag is reduced by a slippery skin or hair covering.

They generate internal body heat from a high metabolic rate, and conserve this heat with layers of insulating fat and, in some cases, fur. Their large size gives them a favorable surface-to-volume ratio -- with less surface area per unit of volume they lose less heat through the skin. This is why there are no marine mammals smaller than a sea otter; a small mammal would lose body heat too rapidly.

Their <u>respiratory systems are modified</u> to collect and retain large quantities of oxygen. The air duct "plumbing" of marine mammals is typically much different from that of land mammals, and the lungs can be more thoroughly emptied before drawing a fresh breath.

They share a number of <u>osmotic adaptations</u> that free them from any requirement for fresh water. Unlike other marine vertebrates, the marine mammals do not have salt-excreting glands or tissues. They swallow little water during feeding (or at any other time), and their skin is impervious to water. This minimal seawater intake, coupled with their kidneys' abilities to excrete a concentrated and highly saline urine, permit them to meet their water needs with the metabolic water derived from the oxidation of food.

Thinking Critically

1. When did the first true animals evolve? What atmospheric changes had to happen before animal life was possible? Are descendants of most of the early forms of animal life represented in the ocean today?

The first animal-like creatures were single-celled organisms. They began to prosper in the ocean during the "oxygen revolution," a time of radical change in the Earth's atmosphere. During the oxygen revolution, between about 2 billion and 400 million years ago, the activity of photosynthetic autotrophs changed the composition of the atmosphere from less than 1% free oxygen to its present oxygen-rich mixture of more than 20%. The growing abundance of free oxygen made it possible for heterotrophs to complete the disassembly of food molecules obtained by eating the autotrophs. Ozone derived from this oxygen blocked most of the sun's dangerous ultraviolet radiation from reaching the Earth's surface, permitting life to survive at the surface of the ocean and, later, on land.

Animals grew in complexity as they became more abundant. Instead of drifting apart after reproduction, some dividing cells stuck together and formed colonies. True animals evolved as these colonies distributed labor among specialized cells, eventually increasing the degree of interdependence between cells within the colony. The colonies ceased to be simple aggregations of individuals and began to take on specific architectures for specific tasks. Multicellular animals had arrived.

If the Burgess Shale (and similar fossil collections) is any indication, comparatively few phyla have lived into the present era. Indeed, most of the animals in these ancient fossils beds are extinct, and may represent "failed experiments" in animal evolution, unique designs that were not suited to later environmental conditions.

2. How can an arthropod grow within a "tailored" shell? How can an animal grow without getting bigger, or get bigger without growing?

The ability to molt and old exoskeleton and generate a new one is the key to arthropod dominance.

We vertebrates grow by steadily adding length to the bones of our *internal* skeleton and bulk to our bodies. An *external* skeleton obviously limits growth and must be shed (molted) at regular intervals. Arthropods do not have a steady growth pattern; instead their external growth progresses in a series of step-like jumps (Figures 15.15) as the animal molts and replaces its exoskeleton. The arthropod grows without getting bigger between these jumps in size. An aquatic arthropod slowly substitutes body mass for water held in the tissues between molts. When molting it suddenly takes on water from outside the body, expanding its tissues without growing in muscle mass. The shell splits and falls away and, through a magnificently orchestrated sequence of glandular secretions, the animal quickly regenerates a new exoskeleton one size larger (Figure 15.17).

3. Are all chordates vertebrates? Are all vertebrates chordates? What distinguishes a vertebrate?

All vertebrates are chordates, but not all chordates are vertebrates. The chordate subphylum Vertebrata, animals with backbones, includes the fishes, amphibians, reptiles, birds, and mammals. Non-vertebrate (invertebrate) chordates include members of the subphylum Urochordata (the tunicates), and the subphylum Cephalochordata (*Amphioxus*). A glance at Table 15.1 will sort out the taxonomy for you.

4. There are seven living classes of vertebrates, but only six are marine. List the seven classes. Which class has no permanent marine representatives? Why nott?

Figure 15.23 lists the seven living classes of vertebrates.

Amphibians have no permanent marine representatives because they depend on the constant flow of water through their skins into their bodies to provide the fluid for the formation of urine to remove nitrogenous wastes. Placing an amphibian in seawater would cause water to flow through the skin in the opposite direction, dehydrating the animal. Their skin is too permeable to permit amphibians to colonize the marine environment.

5. How are odontocete (toothed) whales different from mysticete (baleen) whales? Which are the better known and studied?

Members of the suborder Odontoceti, the toothed whales, are active predators and possess teeth to subdue their prey. Toothed whales have a high brain-weight-to-body-weight ratio, and though much of their "extra" brain tissue is involved in formulating and receiving the sounds on which they depend for feeding and socializing, many researchers believe them to be quite intelligent. Smaller whales in this group include the Orcas (killer whale) and the familiar dolphins and porpoises of oceanarium shows. The largest toothed whale is the 18 meter (60 foot) sperm whale, which can dive to at least 1,140 meters (3,740 feet) in search of the large squids that provide much of its diet.

Toothed whales search for prey using echolocation, the biological equivalent of sonar; they generate sharp clicks and other sounds that bounce off prey species and return to be recognized. Reflected sound is also used to build a "picture" of the animal's environment and to avoid hitting obstacles while swimming at high speed. Odontocete whales are now thought to use sound offensively as well. Recent research indicates that some odontocetes can generate sounds loud enough to stun, debilitate, or even kill their prey. In one experiment dolphins produced clicks as loud as 229 decibels, equivalent to a blasting cap exploding close to the target organism. Sperm whales, it has been calculated, may generate sounds exceeding 260 decibels! (The decibel scale is logarithmic; compare this figure to the 130 decibel noise of a military jet engine at full power 20 feet away!) How this prodigious noise is generated is not yet known, nor do we know how such energy is radiated from the whale without damaging the organs that produce and focus it.

Suborder Mysticeti, the whalebone or baleen whales, have no teeth and are thought to have branched from the line leading to toothed whales early in whale evolution. Filter feeders rather than active predators, these whales subsist primarily on krill, a relatively large shrimp-like crustacean zooplankter obtained in productive polar or sub-polar waters. They do not dive deep, but commonly feed a few meters below the surface. Their mouths contain interleaving triangular plates of bristly horn-like baleen used to filter the zooplankton from great mouthfuls of water or from mud scooped from a shallow seabed. The plankton is concentrated as water is expelled, swept from the baleen plates by the whale's tongue, compressed to wring out as much seawater as possible, and swallowed through a throat not much larger in diameter than a grapefruit. A great blue whale, largest of all animals, requires about 3 metric tons (6,600 pounds) of krill each day during the feeding season; about 1 million Calories per day. The short, efficient food chain from phytoplankton to zooplankton to whale provides the vast quantity of food required for their survival. An extraordinary photograph of a feeding blue whale with its mouth hugely distended with seawater and krill is shown in Figure 15.42.

Toothed whales are the better known and studied of the two suborders.

Thinking Analytically

1. Do you think there are more species of animals on land or in the ocean? What about absolute numbers of animals—are there more animals on land or in the ocean?

Speciation (the mechanism by which new species are created) depends in part on isolation. The ocean flows around land, and is more uniform than land – land appears to offer more opportunities for isolation and thus speciation. Logically, one might assume there would be more species on land than in the ocean, and that idea seems to be holding up as research continues. As for absolute numbers of animals, the ocean is winning. True, there are a bazillion insects on land (and only one known genus with five species in the ocean), but there is no terrestrial equivalent of the planktonic lifestyle. The total number of individual animals in the ocean probably exceeds the total number on land.

2. There are more parasitic species of animals than all other sorts of animals combined. (If this sounds impossible, consider how many animals parasitize any animal you can think about, including ourselves.) Why do you think parasitism is such a runaway success? What are the drawbacks of being a parasite?

Being a parasite means you live in your food. The meal is self-replenishing and self-protecting. External variables are no problem – you're protected from the exigencies of osmotic and thermal shock. You can concentrate on the two most important things in an animal's life: feeding and reproduction. Life is good!

But parasites work harder than you might imagine. They must be exquisitely sensitive to the condition of their host. A lung fluke can't consume too much of a sea lion's lung because a sick sea lion will be unable to eat and will die before the lung fluke can reproduce – the relationship will be a failure in all respects. The biochemical feedback involved is very delicate. Also, reproduction must be timed to allow the parasite's larvae to infect a newly born host at exactly the right time, a host's immune system must be frustrated, its nerve irritation minimized, its senses left unaffected, and its ability to mate retained. This biological high-wire act would seem all but impossible, but (once established) the advantages make the proposition unbeatable. On a per-species basis, the parasitic lifestyle is the dominant animal trophic mechanism.

CHAPTER 16

MARINE COMMUNITIES

Reviewing What You've Learned

1. How does a population differ from a community? A niche from a habitat?

A group of organisms of the same species occupying a specific area is called a <u>population</u>. The many populations of organisms that interact with one another at a particular location comprise a <u>community</u>. A <u>habitat</u> is an organism's "address" within its community, its physical *location*. Each habitat has a degree of environmental uniformity. An organism's <u>niche</u> is its "occupation" within that habitat, its relationship to food and enemies, an expression of what the organism is *doing*.

2. Describe a typical growth curve for a population. What factors influence the shape and eventual height of the curve?

Organisms newly introduced into a favorable environment with no competitors for food or space will reproduce exponentially, tracing a J-shaped population growth curve (as in Figure 16.4). In nature very few populations reproduce at this maximal rate, however, because environmental conditions are rarely ideal and limiting factors in the environment quickly slow the rate of population growth. The sum of the effects of these limiting factors in the environment is called environmental resistance. Environmental resistance causes the actual population growth curve to be lower than the maximum potential growth curve. Figure 16.4 shows the growth rate in number of individuals over time for both potential and actual situations. Note that the curve resulting when limiting factors intrude is S-shaped; it gradually flattens toward an upper limit of the number of individuals in the population. The final number of organisms oscillates around the carrying capacity of the environment for that species: the population size of each species that a community can support indefinitely under a stable set of environmental conditions.

3. Which marine habitat is the most sparsely populated? Why?

Life on the deep ocean floor is more plentiful and obvious than in the bathypelagic water above, but as bottoms go, the density of life at great depths is extremely low. For example, 5,000 grams of living organisms can be found on a square meter of nearshore seafloor, and 200 grams might be recovered from the same area of continental shelf, but less than *1 milligram per square meter* is typical for deep ocean benthic communities! The main difficulty to be overcome is lack of food. There is, of course, no photosynthetic primary productivity at these aphotic depths, and chemosynthetic productivity is usually limited to the richly inhabited rift vents. Organisms on most of the deep bottom must be content with dust-sized scraps falling -- sometimes for miles -- from the productive waters at the ocean surface. Understandably, there is little to eat, and few organisms at the table.

4. Describe the residents of the deep ocean. How do inhabitants of the vent communities differ from other deep floor organisms? What is the source of nutrition for vent communities?

The organisms within deep pelagic and benthic communities share some curious adaptations. Gigantism is a common characteristic: Individuals of representative families tend to be much larger in deep water than related individuals in the shallow ocean. Fragility is also common in the depths. Not only are heavy support structures unnecessary in the calm deep environment, but also the relatively low water pH and high pressure discourage the deposition of calcium and thus skeletal development. Some animals have slender legs or stalks to raise themselves above the sediment, and some come apart like warm gelatin at the slightest touch. Except for its influence on enzyme activity, hydrostatic pressure is not a problem for these animals. They live in balance with the great pressure; their internal pressure is precisely the same as that outside their bodies.

Clustered around hydrothermal vents are dense aggregations of large, previously unknown crabs, clams, sea anemones, shrimp, and unusual worms. Bottom water in the area was laden with hydrogen sulfide, carbon dioxide, and oxygen, on which specialized bacteria were found to live. These bacteria evidently form the base of a food chain that extends to the animals.

The nutrition of the large worms of genus *Riftia* have been the most thoroughly studied. Feeding was something of a puzzle because these animals have no mouth, digestive tract, or anus. The trunks of the worms were found to contain large feeding bodies tightly packed with bacteria similar to those seen in the water and on the bottom near the geothermal vents. The worms' tentacles absorb hydrogen sulfide from the water and transport it to the bacteria, which then use the hydrogen sulfide as an energy source to convert carbon dioxide to organic molecules. The ultimate source of the worms' energy (and the energy of most other residents in this community) is this energy-binding process, called chemosynthesis, which replaces photosynthesis in the world of darkness. Other animals have similar symbiotic/chemosynthetic adaptations.

5. What are the three types of symbioses? Give an example of each.

There are three general types of symbioses: mutualism, commensalism, and parasitism.

In mutualism, as the name implies, both the symbiont and the host -- the larger organism with which the symbiont lives -- benefit from the relationship. True mutualism is rare among marine organisms, but a few examples have been observed. One is the relationship between an anemone fish and its sea anemone. In this symbiosis a small, brightly colored anemone fish nestles within the tentacles of a sea anemone. The mechanism that permits the little fish to do this without being stung is not well understood, but biologists believe the fish gradually desensitizes the stinging cells of the anemone using mucous secretions from its own skin. In return for the anemone's protection the fish feeds the anemone scraps of food and may even lure prey within the anemone's reach.

In commensalism, the symbiont benefits from the association while its host neither benefits nor is harmed. For example, biologists once believed that the association between pilot fish and shark was mutualistic; the pilot fish was thought to guide the shark to a meal and in turn be permitted to dine on the scraps. We now know the pilot fish is only an opportunistic commensal, taking what scraps he can from the shark's meal. (On the other hand, the mutualistic anemone fish/anemone partnership was thought to be commensal before the anemone fish was observed feeding the anemone.)

Parasitism is the most highly evolved and by far the most common symbiotic relationship. The parasite lives in (or on) the host for at least part of its life cycle and obtains food at the host's expense. For obvious reasons, parasites do not usually kill their hosts, but they can seriously affect the host organism by reducing its feeding efficiency, depleting its food reserves, reducing its reproductive potential, lowering its resistance to disease, or otherwise sapping its energy. The host-parasite relationship is finely balanced and extraordinarily delicate. The parasite must in some way be aware of the host's physical condition in order to avoid weakening the host so much that it dies. On the other hand, the parasite must take as much energy from the host as possible in order to ensure its own success.

Thinking Critically

1. What is a limiting factor? Give a few examples.

A physical or biological factor that limits an organism's success in a community is a limiting factor for the organism, a limiting factor that prevents the organism from feeding, growing, reproducing successfully, defending itself, sensing danger, or otherwise functioning successfully. As you may remember from Chapter 13, some tropical fishes die in an unheated home aquarium because their physiology is optimized for a water temperature higher than the interior temperature of a typical house. If their natural tropical environment temporarily dropped to the temperature of a cool room, some of the fishes would become too sluggish to catch fast-moving food or avoid larger predators. Temperature would *limit* their success. Other fishes can adapt relatively easily to changes in temperature, however. Temperature would be a limiting factor to a sensitive tropical fish species, but not to species with a wider temperature tolerance.

The prefixes *steno-* (meaning "narrow") and *eury-* (meaning "wide, broad") are sometimes used to describe those species, populations, or individuals that have narrow or wide tolerance to specific factors. Figure 16.2 provides an idealized look at the tolerance of organisms to a varying physical factor such as temperature.

An animal or plant is almost never exposed to fluctuations of only one physical or biological factor in its environment. An organism may die if subjected to changes in several environmental factors at once, even though each individual change is within its range of tolerance. Even small changes in temperature or salinity, survivable in themselves, might prove lethal if a particular species of tropical fish were exposed simultaneously to both.

Can you think of more examples?

2. In what ways can members of the same population compete with one another? How might members of different populations compete? Contrast the results of these kinds of competition.

When members of the <u>same</u> population (all members of the same species) compete with each other, some individuals will be larger, stronger, or more adept at gathering food, avoiding enemies, or mating. These animals tend to prosper, forcing their less successful relatives to emigrate, fight, or die in the course of competition. The most successful organisms have the most surviving offspring, so useful inheritable variations are passed along in greater quantity to the next generation. As we saw in Chapter 13, this kind of competition continually fine-tunes a population to its environment.

When members of <u>different</u> populations compete, one population may be so successful in its "job" that it eliminates competing populations. In a stable community, two populations cannot occupy the same niche for long. Eventually the more effective competitor overwhelms the less effective one. Extinction from this kind of head-to-head competition is probably uncommon, but restriction of a population because of competition between species is not.

3. What factors influence the distribution of organisms within a community? How are these distributions described? Why is random distribution so rare?

Benthic organisms live on or in the ocean bottom. Some benthic creatures spend their lives buried in sediment, others rarely touch the solid seabed; most attach to, crawl over, swim next to, or otherwise interact with the ocean bottom continuously throughout their lives. Their distribution through space is determined by their needs and by the nature of their interactions with their environment.

The most common pattern for distribution of benthic organisms is small patchy aggregations, or clumps. Clumped distribution occurs when conditions for growth are optimal in small areas because of physical protection (in cracks in an intertidal rock), nutrient concentration (near a dead body lying on the bottom), initial dispersal (near the position of a parent), or social interaction. A random distribution implies that the position of *one* organism in a bottom community in no way influences the position of *other* organisms in the same community. A truly random distribution indicates that conditions are precisely the same throughout the habitat, an extremely unlikely situation except possibly in the unvarying benthic communities of abyssal plains. Uniform distribution with equal space between individuals, such as the arrangement of trees we see in orchards, is the rarest natural pattern of all.

4. What problems confront the inhabitants of the intertidal zone? How do you explain the richness of the intertidal zone in spite of these rigors?

The problems of living in the intertidal zone are formidable. The tide rises and falls, alternately drenching and drying out the animals and plants. Wave shock, the powerful force of crashing waves, tears at the structures and underpinnings of the residents. Temperature can change rapidly as cold water hits warm shells, or as the sun shines directly on newly exposed organisms. In high latitudes ice grinds against the shoreline, and in the tropics intense sunlight bakes the rocks. Predators and grazers from the ocean visit the area at high tide, and those from land have access at low tide. Too much fresh water can osmotically shock the occupants during storms. Annual movement of sediment onshore and offshore can cover and uncover habitats.

In spite of these rigors, the richness, productivity, and diversity of the intertidal rocky community -- especially in the world's temperate zones -- is matched by very few other places. There is intense competition for space. One reason for the great diversity and success of organisms in the rocky intertidal zone is the large quantity of food available. The junction between land and ocean is a natural sink for living and once-living material. The crashing of surf and strong tidal currents keep nutrients stirred and ensure a high concentration of dissolved gases to support a rich population of autotrophs. Minerals dissolved in water running off the land serve as nutrients for the inhabitants of the intertidal zone as well as for plankton in the area. Many of the larval forms and adult organisms of the intertidal community depend on plankton as their primary food source.

In rocky intertidal zones, another reason for the success of organisms is the large number of habitats and niches available occupation (see Figure 16.6). The habitats of intertidal animals and plants vary from hot, high, salty splash pools to cool, dark crevices. These spaces provide hiding places, quiet places to rest, attachment sites, jumping-off spots, cracks from which to peer to obtain a surprise meal, footing from which to launch a sneak attack, secluded mating nooks, or darkness to shield a retreat.

In contrast, sandy beaches are ecological nightmares. Sand itself is the key problem. Many sand grains have sharp pointed edges, so rushing water turns the beach surface into a blizzard of abrasive particles. Jagged grit works its way into soft tissues and wears away protective shells. A small organism's only real protection is to burrow below the surface, but burrowing is difficult without a firm footing. When the grain size of the beach is small, capillary forces can pin down small animals and prevent them from moving at all. If these organisms are trapped near the sand surface, they may be exposed to predation, to overheating or freezing, to osmotic shock from rain, or to crushing as heavy animals walk or slide on the beach. As if this weren't enough, those that survive must contend with the difficulty of separating food from swirling sand and the dangers of leaving telltale signs of their position for predators or being excavated by crashing waves. A few can run for their lives -- some larger beach-dwelling crabs depend on their good eyesight and sprinting ability to outrun onrushing waves. To these horrors must be added all the usual problems of intertidal life discussed above. Not surprisingly, *very* few species have adapted to wave-swept sandy beaches!

5. Why must the host-parasite relationship be so finely balanced? What would be the result of an imbalance?

The parasite will take as much energy from the host as possible, but must not cause the host to die or become so debilitated that predators can catch and eat the host (and the parasite)! It is a very fine balancing act, and one requiring exquisite sensitivity on the part of the parasite to the condition of the host. Parasites and hosts evolve together, one adapting to the other. As you saw in Questions from Students #2 at the end of the chapter, if a parasite enters the wrong host -- a host for which it is not adapted -- the results can be catastrophic for both parties.

Thinking Analytically

1. Why would larvae dispersing from hydrothermal vent communities need the assistance of whale-fall "stepping stones" to colonize new vents? (Hint: Do you think hydrothermal vents steady, relatively stable phenomena?)

Hydrothermal vents come and go, and so must their inhabitants. The larval forms of members of vent communities are a hardy lot, but their development into adults cannot be retarded indefinitely, and they must eventually settle onto a substrate in the vicinity of suitable nutrients. If the distance between active vents is too great, or if vents become inactive in a geological area for relatively long periods of time, the widely dispersed and nutritionally rich whale fall zones will act as temporary camps. Larvae launched from these outposts over the decades (centuries?) have a chance of colonizing newly opened vents.

2. What implications do you think the discovery of living things in deep rocks might have for discovering life on other planets?

The discovery of bacteria thriving at depths of 3.2 kilometers (2 miles) beneath the sea floor was something of a surprise! Their unusual nature (see page 407) suggests they may be remnants of the first forms of life to evolve on Earth. The evolution of surface organisms occurred in different directions (and apparently more rapidly) than the evolution of deep extremophiles, but it may very well be that the biomass of these deep and simple creatures exceeds – perhaps by an order of magnitude – the biomass of marine and terrestrial surface life.

As I write, two NASA rovers (*Spirit* and *Opportunity*) are exploring the surface of Mars. Data beamed back to headquarters strongly suggests the surface of Mars was once wet with saline fluids. The microscopes on the rovers are not powerful enough to detect structures as small as bacteria, so further research will be needed. What would be the implications if extremophiles (or traces of their past inhabitation) were found on Mars?

3. What percent of the total biosphere do you think deep extremophiles comprise? Which is the dominant mechanism of primary productivity on Earth – photosynthesis or chemosynthesis?

No one knows! Biologists will be working for decades on an answer. We are much more familiar with the surficial organisms – we have studied them for decades and understand their biochemistry and natural history reasonably well. Until perhaps 10 years ago most researchers assumed stories of living things in deep rocks were just that – stories. It was only when exquisite care was taken to prevent contamination of deep samples by bacteria carried down the holes by drilling mechanisms that the majority of biologists became convinced that deep rock communities existed. Now we await census samples and estimates of biomass. I'm not a betting man, but I'd place a US$20 bill on the deep biosphere being at least equal to surficial biology in biomass.

CHAPTER 17

MARINE RESOURCES

Reviewing What You've Learned

1. Distinguish between physical and biological resources, and between renewable and non-renewable resources.

Physical resources result from the deposition, precipitation, or accumulation of useful substances in the ocean or seabed. Most physical resources are mineral deposits, but petroleum and natural gas, mostly remnants of once-living organisms, are included in this category. Fresh water obtained from the ocean is also a physical resource. Biological resources are living animals and plants collected for human use.

Renewable resources are naturally replaced on a seasonal basis by the growth of marine organisms or by other natural processes. Non-renewable resources such as oil, gas, and solid mineral deposits are present in the ocean in fixed amounts and cannot be replenished over time spans as short as human lifetimes.

2. What are the three most valuable physical resources? How does the contribution of each to the world economy compare to the contribution of that resource derived from land?

The three most valuable physical resources are petroleum and <u>natural gas</u>, <u>sand and gravel</u>, and <u>fresh water</u>.

About 35% of the crude oil and 26% of the natural gas produced in 2000 came from the seabed. More than 1 billion metric tons (1.1 billion tons) of sand and gravel valued at more than a half of a billion dollars were mined offshore in 1998. Only about 1% of the world's total sand and gravel production is scraped and dredged from continental shelves each year, but the seafloor supplies about 20% of the sand and gravel used in the island nations of Japan and the United Kingdom. Potable water derived from the ocean makes an insignificant contribution to the total amount of fresh water available to the world's human population. Still, desalination is becoming a big business. More than 1,500 desalination plants are presently operating in dry areas worldwide, producing a total of about 13.3 billion liters (3.5 billion gallons) of fresh water per day.

3. What are the sources of metals mined or extracted (or potentially mined or extracted) from the sea?

Magnesium, the third most abundant dissolved element, precipitates from seawater mainly in the form of magnesium chloride ($MgCl_2$) and magnesium sulfate ($MgSO_4$) salts. Magnesium metal, a strong, lightweight material used in aircraft and structural applications, can be extracted by chemical and electrical means from a concentrated brine of these salts.

The recent discovery of metal-rich sulfides around hydrothermal vents has spurred interest among economists as well as oceanographers. Heated seawater carrying large quantities of metals and sulfur leached from the newly formed crust pours out through vents and fractures. The metals -- mainly zinc, iron, copper, lead, silver, and cadmium -- combine with the sulfur and precipitate from the cooler surrounding water as mounds, coatings, and chimneys. While these deposits are certainly commercial-grade ores, they are neither large nor extensive. Also, they are subject to solution and oxidation on the seafloor and are not likely to be preserved in thick layers for long periods of time. Also, three great pools of hot water have been discovered at a depth of about 2,000 meters (6,600 feet) in the Red Sea. The metals precipitating within these basins produce muds rich in metal sulfides, silicates, and oxides of commercial concentration.

4. Distinguish between mariculture and aquaculture. Does fresh water or seawater aquaculture provide more human food, worldwide? What are the prospects for mariculture?

Aquaculture is the growing or farming of plants and animals in any water environment under controlled conditions. Aquaculture production currently accounts for over one-quarter of all fish (fresh water and marine) consumed by humans. Mariculture is the farming of *marine* organisms, usually in estuaries, bays, or near-shore environments, or in specially designed structures using circulated seawater. Mariculture facilities are sometimes placed near power plants to take advantage of the warm seawater flowing from their cooling condensers.

Worldwide mariculture production is thought to be about one tenth that of fresh water aquaculture. Several species of fish, including plaice and salmon, have been grown commercially, and marine and brackish water fish account for 67% of the total production. Several kinds of edible seaweeds are grown, generating 17% of mariculture production. The balance -- 16% -- comes from crustaceans and mollusks, including shrimp, mussels, oysters, and abalone.

The prospects for mariculture, especially for such "luxury seafoods" as shrimp and oysters, seem very promising.

5. What is a non-extractive resource? Give some examples.

Non-extractive resources are uses of the ocean in place; transportation of people and commodities by sea, recreation, and waste disposal are examples. For more insight into the growing economic importance of the ocean as a non-extractive resource, review Box 17.1

Thinking Critically

1. How are oil and natural gas thought to be formed? How can these substances be extracted from the seabed? Why are the physical characteristics of the surrounding rock important?

Petroleum (oil) is almost always associated with marine sediments, suggesting that the organic substances from which it was formed were once marine. Planktonic organisms or soft-bodied benthic marine animals are the most likely candidates. Their bodies apparently accumulated in quiet basins where the supply of oxygen was low and there were few bottom scavengers. The action of anaerobic bacteria converted the original tissues into simpler, relatively insoluble organic compounds that were probably buried, possibly first by turbidity currents, then later by the continuous fall of sediments from the ocean above. Further conversion of the hydrocarbons by high temperatures and pressures must have taken place at considerable depth, probably 2 kilometers (1.2 miles) or more beneath the surface of the ocean floor. Slow cooking under this thick sedimentary blanket for millions of years completed the chemical changes that produce oil.

If the organic material cooked too long, or at too high a temperature, the mixture turned to methane, the dominant component of natural gas. Deep sedimentary layers are older and hotter than shallow ones, and have higher proportions of natural gas to oil. Very few oil deposits have been found below a depth of 3 kilometers (1.8 miles). Below about 7 kilometers (4.4 miles), only natural gas is found.

Oil is less dense than the surrounding sediments, so it can migrate from its source rock through porous overlying formations. It collects in the pore spaces of reservoir rocks when an impermeable overlying layer prevents further upward migration of the oil (see Figure 17.1). When searching for oil, geologists use sound reflected off subsurface structures to look for the signature combination of layered sediments, depth, and reservoir structure before they drill.

2. What method of ocean energy extraction has been most practical? Which has the greatest potential? Why has the latter not been exploited on a large scale?

Tidal power is the only marine energy source that has been successfully exploited on a large scale. The first major tidal power station was opened in

1966 in France on the estuary of the river Rance (see Figure 11.21) where tidal range reaches a maximum of 13.4 meters (44 feet). Built at a cost of $75,600,000, this 850-meter (2,800 foot) long dam contains 24 turbo-alternators capable of generating 544 million kilowatt-hours of electricity annually. At high tide, seawater flows from the ocean through the generators into the estuary. At low tide the seawater and river water from the estuary flow out through the same generators. Power is generated in both directions. A similar installation is generating power at Passamaquoddy Bay, a part of the Bay of Fundy between Maine and New Brunswick, Canada.

The greatest potential for energy generation in the ocean lies in exploiting the thermal gradient between warm surface water and cold deep water. How might the immense potential of thermal gradients be harnessed? Warm seawater would be pumped into the plant through openings near the ocean surface. This water would pass through heat exchangers and boil liquid ammonia -- a liquid with a very low boiling point -- to a pressurized vapor. This gas would turn a turbine, which would spin an electrical generator, and then pass into another heat exchanger cooled by water pumped from the depths. Here the ammonia would condense into a liquid, creating a vacuum that would draw more ammonia vapor through the turbine. The liquid ammonia would be pumped back to the first heat exchanger to repeat the cycle. The system has been dubbed OTEC for Ocean Thermal Energy Conversion.

Why has this promising technology not been exploited? Mainly because of the very low efficiency of the OTEC process. The efficiency of heat-driven power generators depends on the *difference* in temperature between the hottest part of the system and the coldest. Therefore huge amounts of warm and cold water would have to circulate through the plant. An OTEC plant with the same generating capacity as a single large nuclear power plant would need to process a continuous flow of water equal to five times the average flow of the Mississippi River -- and that estimate doesn't include the power necessary to operate the OTEC plant's massive internal pumps.

3. Does the ocean provide a substantial percentage of all protein needed in human nutrition? Of all animal protein? What is the most valuable biological resource? The fastest growing fishery?

Fish, crustaceans, and mollusks contribute about 14.5% of the total animal protein consumed by humans; fish meal and byproducts included in the diets of animals raised for food account for another 3.5%. About 85% of the annual catch of fish, crustaceans, and mollusks comes from the ocean, the rest from fresh water.

The largest commercial harvest is of the herring and its relatives, which accounts for more than a quarter of the live weight of all living marine resources caught each year. The herring and cod fisheries are presently collapsing. By 1992, the population of cod had dropped to about 1/100th of its original size; 35,000 fishery jobs were lost in 1993 and 1994 in eastern Canada alone.

The fastest growing fishery is that for the small euphausiid crustaceans called krill. The ocean around Antarctica supports tremendous numbers of krill, which eat the abundant diatoms of the surface waters and are in turn eaten by baleen whales. Krill have proliferated as the whale population has declined. The Soviets and Japanese pioneered krill harvesting and processing in the late 1960s.

4. What are the signs of overfishing? How does the fishing industry often respond to these signs? What is the usual result? What is bykill?

Overfishing occurs when a species is taken more rapidly than the breeding stock of that species can generate replacements. Even when faced with evidence that it is depleting a stock and disrupting the equilibrium of a fragile ecosystem, the fishing industry's response is usually to increase the number of boats and develop more efficient techniques for capturing animals in order to maintain profits. The result is commercial extinction, depletion of a resource species to a point where it is no longer profitable to harvest. Bykill is the unintended capture (and death) of non-target organisms. In some fisheries, bykill exceeds intended take!

5. What are the advantages and disadvantages of a proclaimed EEZ? Do you feel the United States was justified in proclaiming its own EEZ separate from the provisions of the 1982 United Nations Convention?

The primary advantage is a sense of predictability in legal matters concerning ocean resources -- the sense that nations know where each other stand. Though signatories may selectively honor or ignore individual provisions of agreed-upon treaties, at least the agreements are based on long negotiation and are groundwork for further debate. The 1982 Draft Convention included provision for territorial waters, an EEZ within which nations hold sovereignty over resources, economic activity, and environmental protection, a recognition of the concept of the high seas, and -- at least in endorsement form -- a recognition of the values of protecting the ocean and preventing marine pollution.

Thinking Analytically

1. Imagine a conversation between the owner of a fishing fleet and a governmental official responsible for managing the fishery. List five talking points that each person would bring to a conference table. What would be the likely outcome of the resulting discussion?

I'll let you come up with the talking points, but I'll take a stab at the outcome of the discussion. The fleet owner would be concerned about his debt. He still owes the bank for his hulls and has salaries and pension payments to meet. More fish means more money and the possibility of a debt-free future. The government official has graphs and studies and data suggesting (not proving absolutely – that's not how science works) that maintaining the catch at present levels will deplete the fishery in three to five years. The next quarterly payment is due in 2 months. Five years is a long time away – anything could happen. Let's go fishing! When the fishery collapses, as it did in eastern Canada in 1992 at a cost of 35,000 jobs, everyone will lose.

Thomas Jefferson once wrote that one cannot "legislate temperance…", that is, we pass laws mandating requiring people to act in the common good. That behavior must be learned, and emanate from within. A sustainable low level fishery is always superior to no fishery at all, but, as we've seen, advocates rarely convince those in commercial power.

The chapter opener for the book's last chapter takes up this idea in a different way.

2. Review the information about primary productivity in Chapter 13 (not chapter 14). Estimate the surface area of a wheat or alfalfa field that would produce energy equivalent to the 90 metric tons of phytoplankton required to produce one gallon of gasoline.

Start by re-reading Question #1 on page 434, and then turn to page 307 (Figure 13.3) in your text. Cropland binds an average of around 1,000 grams of carbon per square meter per year, about 8.3 times the all-ocean average for primary productivity (120 $gC/m^2/yr$). 90 metric tons = 90,000 kilograms = 90,000,000 grams. Divide this by 1,000 grams per square meter, and find 90,000 square meters of cropland, or about 22 acres. Again, premium unleaded sounds like a bargain!

CHAPTER 18
ENVIRONMENTAL CONCERNS

Reviewing What You've Learned

1. What is pollution? What factors determine how dangerous a pollutant is?

We define marine pollution as the introduction into the ocean by humans of substances or energy that change the quality of the water or affect the physical and biological environment.

A pollutant causes damage by interfering directly or indirectly with the biochemical processes of an organism. In most cases, an organism's response to a particular pollutant will depend on its sensitivity to the combination of quantity and toxicity of that pollutant. Some pollutants are toxic to organisms in tiny concentrations. For example, the photosynthetic ability of some species of diatoms is diminished when chlorinated hydrocarbon compounds are present in parts-per-trillion quantities. Other pollutants seem harmless, as when fertilizers flowing from agricultural land stimulate plant growth in estuaries. Still other pollutants may be hazardous to some organisms but not to others. For example, crude oil interferes with the delicate feeding structures of zooplankton and coats the feathers of birds but simultaneously serves as a feast for certain bacteria.

Pollutants also vary in their persistence; some reside in the environment for thousands of years while others last only a few minutes. Some pollutants break down into harmless substances spontaneously or through physical processes (like the shattering of large molecules by sunlight). Sometimes pollutants are removed from the environment through biological activity. For example, some marine organisms escape permanent damage by metabolizing hazardous substances to harmless ones. Indeed, many pollutants are ultimately biodegradable, that is, able to be broken down by natural processes into simpler compounds. Most pollutants resist attack by water, air, sunlight, or living organisms, however, because the synthetic compounds of which they are composed resemble nothing in nature.

2. What is eutrophication? How can "good eating" be hazardous to marine life?

Some dissolved organic substances act as nutrients or fertilizers that speed the growth of marine autotrophs, causing eutrophication. Eutrophication is a set of physical, chemical, and biological changes that take place when excessive nutrients are released into the water. Too much fertility may be as destructive as too little. Eutrophication stimulates the growth of some species to the detriment of others, destroying the natural biological balance of an ocean area. The extra nutrients come from wastewater treatment plants, factory effluent, accelerated soil erosion, or fertilizers spread on land. They usually enter the ocean from river runoff, and are particularly prevalent in estuaries. Eutrophication is occurring at the mouths of almost all the world's rivers.

3. What parts of the marine environment are hardest hit by human activities?

The hardest hit habitats are estuaries, the hugely productive coastal areas at the mouths of rivers where fresh water and seawater meet. Pollutants washing down rivers enter the ocean at estuaries, and estuaries often contain harbors with their potentials for oil spills. As little as 1 part of oil for every 10 million parts of water is enough to seriously affect the reproduction and growth of the most sensitive bay and estuarine species. Some of the estuaries along Alaska's Prince William Sound, site of the 1989 *Exxon Valdez* accident, were covered with oil to a depth of 1 meter (3.3 feet) in places. The spill's effects on the $150-million-a-year salmon, herring, and shrimp fishery will be felt for years to come.

Estuaries and bays along the U.S. Gulf Coast, one of the most polluted bodies of water on Earth, are also being severely stressed. About 40% of the nation's most productive fishing grounds, including its most valuable shrimp beds, are found in the Gulf. Nearly 60% of the Gulf's oyster and shrimp harvesting areas -- about 13,800 square kilometers (5,300 square miles) -- are either permanently closed or have restriction placed on them because of rising concentrations of toxic chemicals and sewage. Half of Galveston Bay, once classed as the second most productive estuary in the United States, is off limits to oyster fishers because of sewage discharges.

4. What synthetic chemicals appear to be causing depletion of the Earth's protective ozone layer? What is the likely result?

The primary culprit appears to be the class of chemicals known as chlorofluorocarbons (CFCs) used as cleaning agents, refrigerants, fire-extinguishing fluids, spray-can propellants, and insulating foams. These substances are converted by the energy of sunlight into compounds that attack and partially deplete the Earth's atmospheric ozone.

This decline in ozone alarms scientists because stratospheric ozone intercepts some of the high-energy ultraviolet radiation coming from the sun. Ultraviolet radiation injures living things by breaking strands of DNA and unfolding protein molecules. Species normally exposed to sunlight have evolved defenses against average amounts of ultraviolet radiation, but increased amounts could overwhelm those defenses. Land plants such as soybeans and rice would be subjected to sunburn that decreases their yields. Even plankton in the uppermost two meters of ocean would be affected -- recent research indicates an alarming decrease in phytoplankton primary productivity of between 6% and 12% in the coastal waters around Antarctica. Our own species would not escape: A 1% decrease in atmospheric ozone would probably be accompanied by a 5% to 7% increase in human skin cancer.

5. How is waste heat dangerous to marine life?

Shoreside electrical generating plants use seawater to cool and condense steam. The seawater is returned to the ocean about 6°C (10°F) warmer, a difference that may overstress marine organisms in the effluent area. Some recent power plant designs minimize environmental impact by pumping colder water from farther offshore, warming it to the temperature of the seawater surrounding the plant site, and then releasing it. This method minimizes impact on the surrounding communities, but it still shocks those eggs, larvae, plankton, and other organisms that are sucked through the power plant with the cooling water.

6. What introduced species affect your area? Are they dangerous? Costly?

Alarmingly, several thousand species are in transit every day in the ballast water of tankers and other ships. Juvenile forms of marine organisms can easily hitch rides across otherwise insurmountable oceanic barriers and set up housekeeping at distant shores. These foreign organisms sometimes outcompete native species and reduce biological diversity in their new habitats. New marine diseases can also be introduced in this way. Even canals and fishery enhancement projects can introduce potentially destabilizing new species.

As noted in the text, the Chinese mitten crab (*Eriocheir*) and a common decorative aquarium seaweed (*Caulerpa*) are good examples of the destruction exotic marine species can cause. Probably brought to the western United States and the Atlantic coast of Europe as larval forms in ships' ballast water, the mitten crab is an energetic crustacean that burrows into riverbanks and levees, causing them to collapse. The aquarium alga may be even more of a menace. It spreads through fragmentation. After escaping into the Mediterranean in 1984, this Caribbean native now chokes more than 4050 hectares (10,000 acres) of ocean floor off Spain, France, Italy, and Croatia. It has recently turned up in southern California bays, where it is outcompeting local organisms and greatly reducing biodiversity.

The problems caused by introduced species will grow. There are about 250 known exotic species in California's San Francisco Bay, the most invaded estuary in the world. What problems are near you? Zebra mussels? Fruit-eating insect larvae? More inadvertent global experimentation.

7. What was the largest oil spill? How does this compare with routine introduction of oil into the marine environment by intentional release and leakage from landfills?

History's largest oil spill occurred at the end of the 1991 Persian Gulf War; 240 million gallons were intentionally spilled, and much more oil caught fire and burned for months. Though most people have heard of Alaska's Exxon Valdez accident (1989), few realize that this was only the 46th largest oil spill, and that each year Americans dump much more used engine oil into storm drains and sewers than was spilled by the grounding of the tanker *Exxon Valdez*.

Thinking Critically

1. Why is refined oil more hazardous to the marine environment than crude oil? Which is spilled more often? What happens to oil after it enters the marine environment?

The refining process removes and breaks the heavier components of crude oil and concentrates the remaining lighter, more biologically active ones. Components added to oil during the refining process also make it more deadly. Spills of refined oil, especially near shore where marine life is abundant, can be more disruptive for longer periods of time than spills of crude oil.

Spills of crude oil are generally larger in volume and more frequent than spills of refined oil. Most components of crude oil do not dissolve easily in water, but those that do can harm the delicate juvenile forms of marine organisms even in minute concentrations. The remaining insoluble components form sticky layers on the surface that prevent free diffusion of gases, clog adult organisms' feeding structures, and decrease the sunlight available for photosynthesis. Even so, crude oil is not highly toxic, and it is biodegradable. Though crude oil spills look terrible and generate great media attention, most forms of marine life in an area recover from the effects of a moderate spill within about five years. For example, the 240 million gallons of light crude oil released into the Persian Gulf during the 1991 Gulf War have dissipated relatively quickly and will probably cause little long-term biological damage.

2. What heavy metals are most toxic? How do these substances enter the ocean? How do they move from the ocean to marine organisms and people?

Among the most dangerous heavy metals being introduced into the ocean are mercury and lead. Human activity releases about five times as much mercury and 17 times as much lead as is derived from natural sources, and incidents of mercury and lead poisoning, major causes of brain damage and behavioral disturbances in children, have increased dramatically over the last two decades. As a dad, grandpa, and lover of things oceanic, I find Box 18.1 profoundly unsettling.

Lead particles from industrial wastes, landfills, and gasoline residue reach the ocean through runoff from land during rains, and the lead concentration in some shallow water bottom feeding species is increasing at an alarming rate. Consumers should be wary of seafood taken near shore in industrialized regions.

3. Few synthetic organic chemicals are dangerous in the very low concentrations in which they enter the ocean. How are these concentrations increased? What can be the outcome when organisms in a marine food chain ingest these substances?

The level of synthetic organic chemicals in seawater is usually very low, but some organisms at higher levels in the food chain can concentrate these toxic substances in their flesh. This biological amplification is especially hazardous to top carnivores in a food web. For example, in the early 1960s California pelicans began producing eggs with thin shells containing less than normal amounts of calcium carbonate. The eggs broke easily, no chicks were hatched, and the nests were eventually abandoned. The pelicans were disappearing. The trail led investigators to DDT. Plankton absorbed DDT from the water; fishes that fed on these microscopic organisms accumulated DDT in their tissues; and the birds that fed on the fishes ingested it, too. The whole food chain was contaminated, but because of biological amplification the top carnivores were most strongly affected. A chemical interaction between DDT and the birds' calcium depositing tissues prevented the formation of proper eggshells. DDT was eventually banned in the United States, and the pelican and osprey populations are recovering.

Biological amplification of other chlorinated hydrocarbons has also affected other species. Polychlorinated biphenyls (PCBs), fluids once widely used to cool and insulate electrical devices and to strengthen wood or concrete, may be responsible for the behavior changes and declining fertility of some populations of seals and sea lions on islands off the California coast. PCBs have also been implicated in a deadly viral epidemic among dolphins in the western Mediterranean.

4. What is the greenhouse effect? Is it always detrimental? What gases contribute to the greenhouse effect? Why do most scientists believe the

Earth's average surface temperature will increase over the next few decades? What may result?

Greenhouse effect, named after the similarity of the phenomenon to the warming of a greenhouse by the sun in winter, is the trapping of the sun's heat in the Earth's atmosphere. A certain amount of greenhouse effect is necessary for life; without it, Earth's average atmospheric temperature would be about -18°C (0°F). Earth has been kept warm by natural greenhouse gases, including methane and carbon dioxide. Recently, however, human demand for quick energy to fuel industrial growth, especially since the beginning of the industrial revolution, has injected unnatural amounts of new carbon dioxide into the atmosphere from the combustion of fossil fuels. Reckless burning of forests and jungles exacerbate the problem. Carbon dioxide is now being produced at a greater rate than it can be absorbed. There has been a 4°C (7°F) rise in global temperature from the end of the last ice age until today. Carbon dioxide and other human-generated greenhouse gases produced since 1880 are thought to be responsible for about 1°C (1.8°F) of that increase. If current models of greenhouse warming are correct, we can expect global temperature to rise another 2.5°C (4.5°F) by the year 2030. A temperature increase of this magnitude would cause water in the ocean to expand; average sea level would rise between 8 and 29 centimeters (3 and 11 inches). Imagine the effect on the harbors, coastal cities, river deltas, and wetlands where one-third of the world's people now live.

5. What is the tragedy of the commons? Do you think Garrett Hardin was right in applying the old idea to modern times? What will you do to minimize your negative impact on the ocean and atmosphere?

Hardin's "tragedy of the commons" applies to us all. We tend to keep for ourselves the positive outcomes of our use of environmental resources, yet spread and share the negatives. We want the convenience and safety of modern technology, but hesitate to confront the wastes generated by its employment. In the past, human civilizations could move somewhere else after polluted an area and stripped it of its resources. Not any more. As Tyler Miller has noted, it is impossible to throw anything away -- there is no "away."

Each of us can make a difference. A moment's careful reflection on the environmental consequences of such decisions as purchasing products, traveling to work or class, using air conditioning and heating, and recycling materials can make a large difference if enough of us are willing to make changes. But we have entered a time of inadvertent global experimentation, and the trials ahead will be interesting, indeed. As I wrote in the book's Afterword: *There is much good in the world. Go and add to it.*

6. How might global warming or a decrease in stratospheric ozone directly affect the ocean?

If present models of greenhouse warming are true, the costs to society will be very large. Remember, the bulk of the world's human population lives near and depends (directly or indirectly) upon oceanic resources. For some insight into the possible magnitude of the problem, consider the damage to the world's seaports if sea level were to increase by 1 meter. Coastal land values? Erosion? And what of different distributions of crops? Disease? Infrastructure? This promises to be a fascinating millennium.

The decline in ozone alarms scientists because stratospheric ozone intercepts some of the high-energy ultraviolet radiation coming from the sun. Ultraviolet radiation injures living things by breaking strands of DNA and unfolding protein molecules. Species normally exposed to sunlight have evolved defenses against average amounts of ultraviolet radiation, but increased amounts could overwhelm those defenses. Land plants such as soybeans and rice would be subjected to sunburn that decreases their yields. Even plankton in the uppermost 2 meters of ocean would be affected; recent research indicates an alarming decrease in phytoplankton primary productivity of between 6% and 12% in the coastal waters around Antarctica.

Reminds me of the old Chinese curse: "May you live in interesting times."

Thinking Analytically

1. The cost of pollution and habitat mismanagement, over time, will be higher than the cost of doing nothing. But the cost now is cheaper. Arguing only from practical standpoints (that is, avoiding an appeal to emotion), how could you convince the executive board of a first-world industrial corporation dependent on an ocean resource to reduce or eliminate the negative effects of its activities.

Take the 1993 and 1994 bankruptcy proceeding records for the Canadian province of Newfoundland to the meeting. More than 18,000 people in this marine-dependent province lost their jobs, homes, and possessions in the complete collapse in 1992 of the cod fishery. What could a bank do with a repossessed fishing boat? Crime rose, tourism plummeted. Banks failed, and the cascade of fiscal calamity was felt to Ottawa and beyond. "Do you want this to happen to your State?" It might work; it might not.

2. How would you convince the board of a corporation in a developing country (say, China)?

China is an interesting case. The country is rapidly moving from a centrally governed economy to an entrepreneurial one. Some areas (Hong Kong, Macau, and the special economic zones surrounding these places) are more autonomous than others, but there is strong national coordination. Unlike most western countries, China tends to take a very long view of development and is planning infrastructure, energy economics, and industrial growth with a view toward long-term sustainability. China has tremendous challenges ahead in cleaning up pollution, managing population, and providing for agriculture, of course, but one can imagine that if the central policy makers were convinced that fisheries were being depleted, effective steps would be taken to control overfishing. That this has yet to occur is clearly evident in Figure 17.18.

3. If a pollutant has effects at a dilution of one part-per-billion by weight, how much seawater is contaminated by the release of one metric ton of the material into a bay? Consider either Chesapeake Bay or San Francisco Bay in your calculations.

One used to hear: "The solution to pollution is dilution." Not so. Research has shown that astonishingly small concentrations of some pollutants are able to cause direct damage or damage after concentration through food chains. One metric ton = 1,000 kilograms = 1,000,000 grams. *One* gram of a pollutant could contaminate about one billion grams of seawater = 1,000 metric tons. *One metric ton* of a pollutant could therefore contaminate one billion metric tons of seawater, roughly one cubic kilometer of seawater. The area of San Francisco Bay is around 1,165 square kilometers. Assuming an average depth of 100 meters, one metric ton of pollutant could contaminate nearly a tenth of the water in San Francisco Bay!

4. You and your wife spend a two-week vacation in Hawaii during which you eat top marine carnivores (bluefin and albacore tuna, swordfish, marlin) for most lunches, and each evening for dinner. A month after your return, you and your wife are pleased to discover you're pregnant. How concerned should you be about the health of your child?

Tough call. First, check the Monterey Bay Aquarium's Seafood Watch website:

http://www.mbayaq.org/cr/cr_seafoodwatch/sfw_aboutsfw.asp

Note that many of these animals are on the Aquarium's "avoid" list, mainly because of heavy metal contamination and/or maricultural practices that damage the local environments. Your immediate concern would be mercury toxicity. Mercury in large quantities can be neurologically devastating (see again Box 18.1). Mercury enters marine food chains from anthropogenic pollution (as in Minamata Bay), natural oceanic sources, and the settling of ash from coal-burning power plants. The good news about Hawaiian fish is that they have less of a burden of contaminants than fishes caught of the U.S. east coast (remember prevailing winds). The important thing for your wife: Through the balance of the pregnancy, don't eat any more tilefish, albacore tuna, or swordfish – all serious potential sources of mercury pollution. When the child is born, continue avoid those fishes if breast feeding. If bottle feeding, it is probably OK to eat seafood once a week, but tilefish, albacore tuna, or swordfish should still be avoided. As for you, remember that even small concentrations of heavy metal have been shown to affect sperm production and motility. For more information, I urge you both to read the governmental guidelines regularly updated in the websites listed through the appropriate sections in Chapter 18.